A NATURALIST
IN THE AMAZON

A NATURALIST
IN THE AMAZON

*The Journals & Writings of
Henry Walter Bates*

Smithsonian Books

Washington, DC

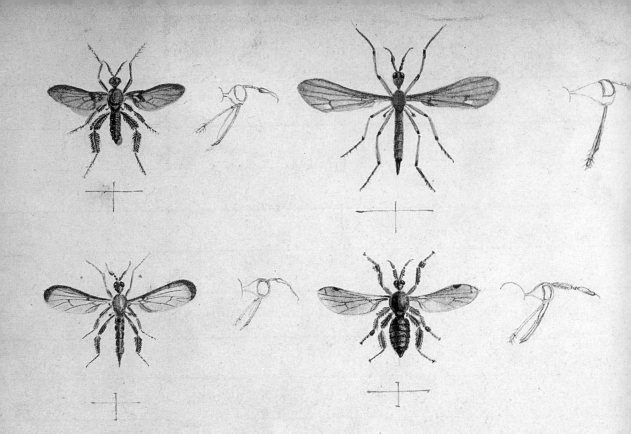

PUBLISHER'S NOTE

Henry Walter Bates, famous for his theory of Batesian mimicry and his publication *The Naturalist on the River Amazons* that was first published in 1863 and is still in print today, also wrote and illustrated the two journals featured in this book. Bates's rarely reproduced journals, together titled *Insect Fauna of the Amazon Valley*, reside in the Natural History Museum, London. The journal sections included herein are the most interesting ones; both journals run to hundreds of pages. The pages are reproduced at actual size, and the border area at the edges varies because the journals differ slightly in size.

Excerpts from *The Naturalist on the River Amazons*, taken from the original two-volume edition published by John Murray, introduce the journal pages. The first section of excerpts is drawn from volume I, and the second and third sections are taken from volume II. The excerpts focus on Bates's descriptions and observations of the natural history he encountered in the Amazon. The subheadings represent the beginning of a new excerpt, and ellipses indicate where some text has been omitted.

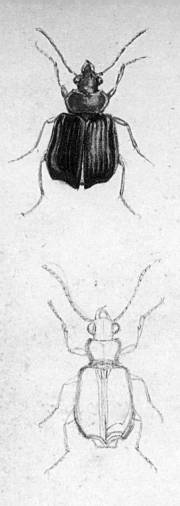
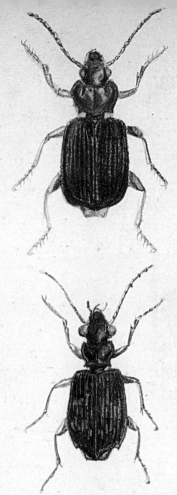

CONTENTS

Photograph of Henry Bates when he was
assistant secretary of the Royal Geographical
Society, a post he took up in 1864, five years
after returning from the Amazon.

INTRODUCTION

HENRY WALTER BATES was born in Leicester on 8th February, 1825. The oldest son of a hosier, he left school at the age of 13 and became an apprentice at a hosiery manufacturer in the expectation of joining his father's trade. From an early age he showed a strong interest in natural history, particularly entomology, and his enthusiasm was further fired in 1844 when he met and became friends with Alfred Russel Wallace, who was, at the time, teaching English at the Collegiate School in Leicester.

United by a shared interest in reading and collecting beetles, the two men were particularly inspired by Alexander von Humboldt's accounts of his travels in South America and Charles Darwin's travel diary *The Voyage of the Beagle*. Together, they decided to devote their lives to their hobby and fixed on Amazonian Brazil as their first destination to 'gather facts towards solving the problem of [the] origin of species'.

Unlike Darwin, Bates and Wallace were not independently wealthy. The only way they could finance their travels was by collecting specimens and sending them back to England to supply the seemingly insatiable mid-nineteenth-century appetite for new and exotic species from previously unexplored parts of the world. Before setting off on their travels, the two men parted company for a couple of years when Wallace went back to Wales to sort out his brother's surveying business. Surveying was very profitable at the time, due to the rapidly extending railway system, and it enabled

Wallace to accumulate sufficient funds for him and Bates to fulfil their dreams and set off for South America.

With assurance from the authorities at the British Museum that any specimens they collected would find a ready market, and with an arrangement in place with an agent, Samuel Stevens of the then famous Steven's Auctions, Bates and Wallace sailed from Liverpool in April 1848 to Pará (present day Belém), near the mouth of the Amazon. It is difficult today to appreciate the enormity of making such a trip in the 1840s. Modern transport and communications have not only made travel far easier, but our increased understanding of tropical diseases and their transmission, prevention and treatment have made everything much safer.

Apart from a journey of about 160 kilometres (100 miles) down the Rio Tocantins, Bates and Wallace spent the first year of their travels in Pará, learning the local customs and what was to become their livelihood, animal collecting. The diversity of animals, and particularly insects, in the forests and around the rivers surpassed all their expectations. In their first two months of collecting Wallace could report to Samuel Stevens that they had collected no less than '553 species of Lepidoptera ... 450 beetles, and 400 of other orders'.

Following their joint expedition along the Tocantins, Bates and Wallace generally journeyed separately, presumably because both were collecting to finance further exploration and so it made no sense to be doing it in the same place. Wallace stayed in South America for four years in all, journeying in the upper Amazonian region and concentrating particularly on the Rio Negro and the Rio Vaupés. He returned to England in 1852 and was lucky to survive the journey home as the brigantine he was travelling in caught fire and he was forced to watch from a lifeboat as his painstakingly collected specimens sank to the bottom of the ocean. Next he headed to Southeast Asia, and it was from there that he assured his place in history through his marvellous collecting and writing, in particular his essay entitled *On the Tendency of Varieties to Depart Indefinitely from the Original Type*. It outlined an explanation for how species changed through time and so cemented his position as one of the great thinkers of the age.

Table of Average Temperatures — Climate of Parà

1845	Sunrise	Noon	8 P.M.	Extremes
January	76.10	85.17	79.81	90 – 74 (80.50)
February	74.86	83.01	78.28	90 – 73
March	75.68	84.16	78.93	91 – 74
April	76.24	85.76	79.55	90 – 73
May	76.64	87.03	80.58	90 – 75
June	76.31	87.07	80.61	90 – 75
July	75.74	87.52	80.56	90 – 74
August	75.93	86.97	80.77	90 – 74
September	75.93	87.77	81.57	90 – 74
October	76.13	87.37	81.83	92 – 74
November	76.27	88.10	82.17	92 – 74
December	76.03	86.74	80.77	91 – 75

Mean of year 81.33

Days without rain 94

„ with Do 271

1846				
January	76.30	84.67	79.40	88 – 75 (80.10)
February	75.80	83.80	79.20	89 – 74
March	75.19	81.35	78.4	89 – 73
April	76.07	84.97	79.63	89 – 75
May	76.74	87.03	80.74	91 – 76

1846	Sunrise	Noon	8 P.M.	Extremes
June	77.03	87.47	81.97	91 – 76
July	77.29	90.32	83.19	95 – 73
August	76.16	88.71	81.77	94 – 75
September	76.10	87.93	81.53	92 – 74
October	76.07	87.79	81.61	92 – 75
November	76.33	87.60	81.53	93 – 75
December	76.58	87.06	82.35	92 – 76

Mean of the year 81.41

Days without rain 100 with Do 265

1847				
January	75.51	83.58	79.57	90 – 74
February	75.00	84.64	78.36	90 – 73
March	75.60	82.00	78.70	90 – 74
April	75.70	82.73	78.43	90 – 74
May	75.71	84.80	78.23	90 – 74
June	75.40	86.71	79.41	90 – 74
July	75.54	87.17	79.88	93 – 74
August	76.07	88.33	80.35	92 – 74
September	73.51	87.93	78.72	91 – 70
October	75.84	87.23	80.19	74 – 92
November	76.14	87.28	81.64	90 – 75
December	75.79	86.14	80.07	92 – 74

Mean of year 80.66

Days without rain 95 with Do 260

A page from Henry Bates's first journal, in
which he meticulously recorded the average
temperatures and climate of Pará.

One of the illustrations that Bates drew on a loose piece of paper while on his travels and then pinned into his journal.

Bates, by contrast, stayed on in the Amazon for a total of 11 years, and he was careful to send his specimens back in smaller shipments to avoid them suffering a fate similar to that of Wallace's. For the most part he travelled alone, collecting with enormous dedication, meticulously recording his captures and the relative productivity of different sites. He lived frugally, travelling mostly by water, writing and drawing in his journals as he went. During the course of those years he contracted both malaria and yellow fever, and his health never fully recovered even after he returned to England.

Although an entomologist at heart, Bates did not let that limit him when it came to collecting and, in addition to 14,000 species of insects, he collected some 712 species of mammals, reptiles, birds, fishes and molluscs, employing local people to help him along the way. But it was not just Bates's collecting that marked him out as being so exceptional. The length and intensity of his exposure to the forests, together with his generally solitary

lifestyle, allowed him to make detailed observations of the kind hardly possible today. He learned the language and customs of many of the local people, including then declining and now vanished tribes of Amazonian Indians, and made detailed observations of the forests themselves and the insects and birds that he was most interested in collecting.

It was the time, inclination and ability to observe that led him to the discovery for which he is best known today, and which bears his name – Batesian Mimicry. Bates noticed early on, while collecting butterflies, that certain brightly coloured species of the genus then called *Heliconius* were slow flying and conspicuous compared to other butterflies. But, in spite of their apparent vulnerability to predation, they were seldom eaten by the numerous birds and other insect-eating animals. He concluded, partly through observation, that this was because they sequestered toxic chemicals that made them poisonous, bitter or unpalatable to predators, which, after one or two bad experiences, learned to avoid a specific colour pattern. He also observed similar-looking butterflies, in some cases belonging to different genera and families, that closely matched the colouration of the toxic species and often flew together with them, although apparently not toxic themselves. He inferred that these non-toxic species, the 'mimics', were obtaining protection by their resemblance to the more abundant, toxic species, called 'models', and supposed that if a harmless butterfly were to produce a colour form that only slightly resembled a toxic model, it would gain a slight degree of protection and would be more likely to survive and reproduce. This effect, concentrated over hundreds of generations, could eventually produce a harmless mimic so closely resembling the toxic model that an entomologist would need to collect and examine the specimen to be certain what it was – which was exactly what Bates was doing when he became aware of the phenomenon. The closest mimics resembled their models not only in colour but in behaviour, flying at similar speeds, heights, and times of day; even though in some cases they were quite unrelated, they could be near indistinguishable on the wing.

By the time Bates arrived back in England in 1859, at only 34 years old, he had already made something of a name for himself. As well as his

specimens, he had sent back manuscripts of scientific notes and papers that Stevens had submitted to journals on his behalf. On his return he began work on his magnum opus, a description of his travels and the scientific and other observations he had made during them, drawing on the journals and other notes that he had written while he was there. He sent draft chapters to Charles Darwin for him to review, and Darwin encouraged him to continue, recommending the book to John Murray, who had published his *On the Origin of Species* in 1859.

The Naturalist on the River Amazons was published in 1863 to wide acclaim, with Darwin himself describing it as 'the best book of Natural History Travels ever published in England.' The two-volume first edition of 1,250 copies sold out in a few months, and his publisher then persuaded Bates to produce an abridged single volume edition, with most of the technical description removed. This edition was then reprinted many times, and it was only in 1892 that the original unabridged edition was published again, together with a memoir of Bates by Edward Codd.

Bates sought employment at the British Museum, but this was blocked by certain established taxonomists who took exception to the way his book and academic papers supported Darwinian evolutionary theory. He was, however, appointed Assistant Secretary at the Royal Geographical Society in 1864, and it was a position

A loose page with illustrations of Lepidoptera that Bates pinned into his journal.

The receipt for the sale of Bates's journals to John James Joicey in 1917. They had been sold to Dulau & Co. directly by the Bates family. The Natural History Museum acquired them from Joicey in 1933, when he went bankrupt and had to sell off his natural history collections.

he held for 27 years. During those years he also served as President of the Entomological Society for two terms and was elected a Fellow of the Linnaean Society in 1871 and a Fellow of the Royal Society in 1881.

On his return from the Amazon, Bates had sold a large part of his collections to the British Museum, which were then transferred to the Natural History Museum. Almost 75 years later two journals that he had written and illustrated during his 11 years in the Amazon were added to the collection. These were bought by the Museum from amateur entomologist John James Joicey, who had to sell off his extensive natural history collections when he went bankrupt in the early 1930s. Joicey had purchased the journals in 1917 from a book dealer to whom the Bates family had sold them.

This book is a celebration of Bates's published and unpublished works. It includes three sections of selected excerpts from the first edition of *The Naturalist on the River Amazons*, the first from his time in Pará and the lower Amazons, the second from his travels further inland from Santarém up to his arrival at Ega, and the third from his time at Ega and the upper Amazons and his eventual return to Pará. In between these sections are full-size reproductions of selected pages from his two handwritten journals. The first journal features notes and drawings of a wide range of different insects, while the second is focused more on Lepidoptera. Together, the excerpts and pages from his journals provide a handsome testament to the achievements of one of the most outstanding naturalist-explorers of his time.

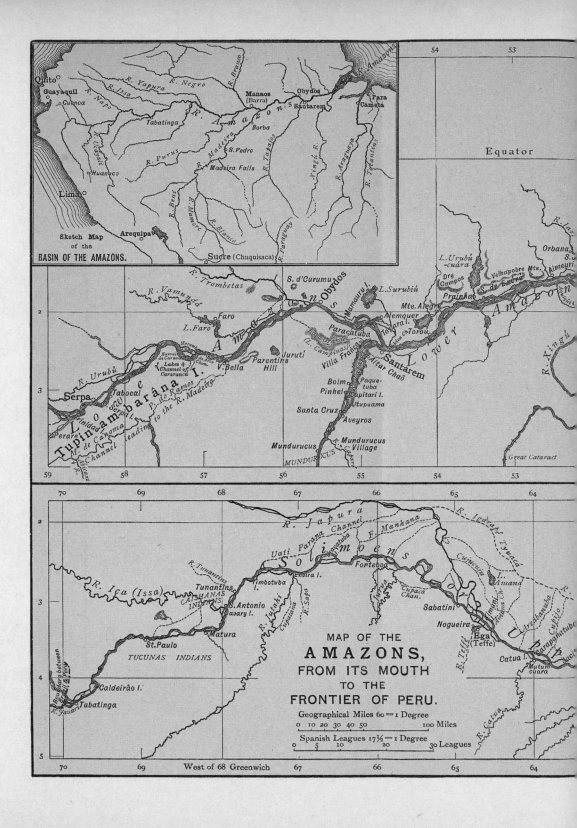

Sketch Map of the BASIN OF THE AMAZONS.

Quito
Guayaquil
Cuenca
R. Yapura
R. Isso
R. Napo
R. Branco
R. Negro
Manaos (Barra)
Obydos
Santarem
Para
Cametá
R. Amazons
Equator
Tabatinga
Borba
R. Ucayali
R. Amazons
Huanuco
R. Purus
R. Madeira
S. Pedro
R. Tapajos
Xingu R.
R. Araguaya
Lima
Madeira Falls
R. Beni
R. Mamoré
R. Blanco
R. Tocantins
Arequipa
Sucre (Chuquisaca)
R. Paraguay

R. Vamundá
R. Trombetas
S. d'Curumu
Obydos
L. Urubú-cuára
Orbana
Almeyri
Faro
Mamauru I.
L. Surubiú
Dry Campos
Velhapobre Mts.
S. de Erere
R. Ja
L. Faro
Mte. Alegre
Prainha
Nemquer
Amazon
Parachtuba
Iguira I.
Torou.
Lower
Serpa
Tabocal
Morena
Barreira de Cararú
da Serra Chan.
Lakes & Channel of Cararuca
Juruti
U. Campinas
Andes
Villa Franca
Altar Chaô
Santarem
R. Xingu
R. Urubú
V. Bella
Parentins Hill
Boim
Paquetuba
Trinidad
Serpal.
P. do Ramos
Pinhel
Capitari I.
Perare
M. de Cáuoma
Channel leading to the R. Madeira
Santa Cruz
Itupuama
Abacaxi R.
Mundurucus
Mundurucus Village
Aveyros
Tupinambarâna I.
R. Cararu
MUNDURUCUS
Great Cataract

R. Japura
R. Igarapi Tyuaca
Uati Parana Channel
Paratuba
F. Manhana
Soquetuba
Solimoens
R. Tunantins
Cuiucutu
L. Amana
R. Ica (Issa)
Tunantins
Timbotuba
Envira I.
Forteboa
Inirida
Cupaca Chan.
CAISHANAS INDIANS
S. Antonio
Javary I.
R. Jutahi
Cupatiana
R. Sapa
Sabatini
R. Teffe
Ymini Ch.
Arvhaia Ch.
Matura
Nogueira
U.
St. Paulo
Ega (Teffe)
Areghanaha
TUCUNAS INDIANS
Catua
Catua I.
Mutum-cuara
Caldeirão I.
Boundary between Brazil & Peru
R. Jauari
Tabatinga
R. Catua

MAP OF THE AMAZONS, FROM ITS MOUTH TO THE FRONTIER OF PERU.

Geographical Miles 60 = 1 Degree

0 10 20 30 40 50   100 Miles

Spanish Leagues 17½ = 1 Degree

0   5   10   20   30 Leagues

West of 68 Greenwich

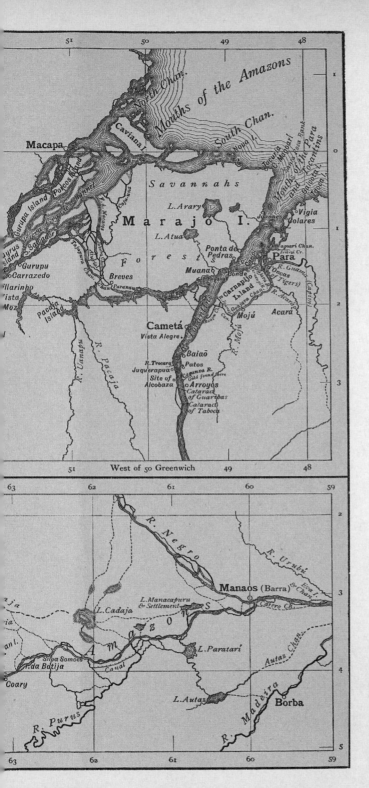

This map accompanied the first edition of *The Naturalist on the River Amazons*. Inset top left is a sketch of the Amazon basin in its entirety; the map at the bottom is a continuation of the map above it, showing the course of the Amazon and its tributaries from its mouth to the frontier of Peru.

Adventure with curl-crested toucans

The Naturalist on the River Amazons

PREFACE

IN THE AUTUMN OF 1847 Mr. A. R. Wallace, who has since acquired wide fame in connection with the Darwinian theory of Natural Selection, proposed to me a joint expedition to the River Amazons, for the purpose of exploring the Natural History of its banks; the plan being to make for ourselves a collection of objects, dispose of the duplicates in London to pay expenses, and gather facts, as Mr. Wallace expressed it in one of his letters, "towards solving the problem of the origin of species," a subject on which we had conversed and corresponded much together. We met in London, early in the following year, to study South American animals and plants at the principal collections; and in the month of April, as related in the following narrative, commenced our journey. My companion left the country at the end of four years; and, on arriving in England, published a narrative of his voyage, under the title of *Travels on the Amazons and Rio Negro*. I remained seven years longer, returning home in July, 1859; and having taken, after the first two years, a different route from that of my friend, an account of my separate travels and experiences seems not an inappropriate offering to the public.

When I first arrived in England, being much depressed in health and spirits after eleven years' residence within four degrees of the equator, the last three of which were spent in the wild country 1,400 miles from the sea-coast, I saw little prospect of ever giving my narrative to the world; and

indeed, after two years had elapsed, had almost abandoned the intention of doing so. At that date I became acquainted with Mr. Darwin, who, having formed a flattering opinion of my ability for the task, strongly urged me to write a book, and reminded me of it months afterwards, when, after having made a commencement, my half-formed resolution began to give way. Under this encouragement the arduous task is at length accomplished. It seems necessary to make this statement, as it explains why so long a time has intervened between my arrival in England and the publication of my book.

The collections that I made during the whole eleven years were sent, at intervals of a few months, to London for distribution, except a set of species reserved for my own study, which remained with me, and always accompanied me in my longer excursions. With the exception of a few living plants and specimens in illustration of Economical and Medicinal Botany, these collections embraced only the Zoological productions of the region. The following is an approximative enumeration of the total number of species of the various classes which I obtained: —

Mammals	.	.	.	52
Birds	.	.	.	360
Reptiles	.	.	.	140
Fishes	.	.	.	120
Insects	.	.	.	14,000
Molluscs .	.	.	.	35
Zoophytes	.	.	.	5
				14,712

The part of the Amazons region where I resided longest being unexplored country to the Naturalist, no less than 8,000 of the species here enumerated were new to science, and these are now occupying the busy pens of a number of learned men in different parts of Europe to describe them.

PARÁ TO OBYDOS

I EMBARKED AT LIVERPOOL, with Mr. Wallace, in a small trading vessel, on the 26th of April, 1848; and, after a swift passage from the Irish Channel to the equator, arrived, on the 26th of May, off Salinas. This is the pilot-station for vessels bound to Pará, the only port of entry to the vast region watered by the Amazons. It is a small village, formerly a missionary settlement of the Jesuits, situated a few miles to the eastward of the Pará river. Here the ship anchored in the open sea, at a distance of six miles from the shore, the shallowness of the water far out around the mouth of the great river not permitting in safety a nearer approach; and the signal was hoisted for a pilot. It was with deep interest that my companion and myself, both now about to see and examine the beauties of a tropical country for the first time, gazed on the land, where I, at least, eventually spent eleven of the best years of my life.

OUR FIRST WALK ASHORE

The impressions received during this first walk can never wholly fade from my mind....

As we continued our walk the brief twilight commenced, and the sounds of multifarious life came from the vegetation around. The whirring of cicadas; the shrill stridulation of a vast number and variety of field crickets

and grasshoppers, — each species sounding its peculiar note; the plaintive hooting of tree frogs — all blended together in one continuous ringing sound, — the audible expression of the teeming profusion of Nature. As night came on, many species of frogs and toads in the marshy places joined in the chorus: their croaking and drumming, far louder than anything I had before heard in the same line, being added to the other noises, created an almost deafening din. This uproar of life, I afterwards found, never wholly ceased, night or day: in course of time I became, like other residents, accustomed to it. It is, however, one of the peculiarities of a tropical — at least, a Brazilian — climate which is most likely to surprise a stranger. After my return to England the death-like stillness of summer days in the country appeared to me as strange as the ringing uproar did on my first arrival at Pará.

COLOUR OF TROPICAL INSECTS

It is a notion generally entertained that the superior size and beauty of tropical insects and birds are immediately due to the physical conditions of a tropical climate, or are in some way directly connected with them. I think this notion is an incorrect one, and that there are other causes more powerful than climatal conditions which affect the dress of species. To test this we ought to compare the members of those genera which are common to two regions; say, to northern Europe and equinoctial America, and ascertain which climate produces the largest and most beautifully coloured species. We should thus see the supposed effects of climate on nearly allied congeners, that is, creatures very similarly organised. In the first family of the order Coleoptera, for instance, the tiger-beetles (Cicindelidae), there is one genus, *Cicindela*, common to the two regions. The species found in the Amazons Valley have precisely the same habits as their English brethren, running and flying over sandy soils in the bright sunshine. About the same number is found in each of the two countries: but all the Amazonian species are far smaller in size and more obscure in colour than those inhabiting northern Europe; none being at all equal in these respects to the common English *Cicindela campestris*, the handsome light-green tiger-beetle, spotted with

white, which is familiar to country residents of Natural History tastes in
most parts of England. In butterflies I find there are eight genera common
to the two regions we are thus pitting against each other. Of these, three
only (*Papilio*, *Pieris* and *Thecla*) are represented by handsomer species in
Amazonia than in northern Europe. Three others (*Lycgena*, *Melitaea* and
Apatura) yield far more beautiful and larger forms in England than in the
Amazonian plains; as to the remaining two (*Pamphila* and *Pyrgus*) there is
scarcely any difference. There is another and hitherto neglected fact which
I would strongly press upon those who are interested in these subjects. This
is, that it is almost always the males only which are beautiful in colours. The
brilliant dress is rarely worn by both sexes of the same species: if climate
has any direct influence in this matter, why have not both sexes felt its
effects, and why are the males of genera living under our gloomy English
skies adorned with bright colours?

Leaf-Carrying Ant

Another far more interesting species was the Saüba (*Oecodoma cephalotes*).
This ant is seen everywhere about the suburbs, marching to and fro in broad
columns. From its habit of despoiling the most valuable cultivated trees of
their foliage, it is a great scourge to the Brazilians. In some districts it is so
abundant that agriculture is almost impossible, and everywhere complaints
are heard of the terrible pest.

The workers of this species are of three orders, and vary in size
from two to seven lines; some idea of them may be obtained from the

Saüba or Leaf-carrying ant
1. Worker-minor
2. Worker-major
3. Subterranean worker

accompanying wood-cut. The true working-class of a colony is formed
by the small-sized order of workers, the worker-minors as they are called
(Fig. 1). The two other kinds, whose functions, as we shall see, are not yet
properly understood, have enormously swollen and massive heads; in one
(Fig. 2), the head is highly polished; in the other (Fig. 3), it is opaque and
hairy. The worker-minors vary greatly in size, some being double the bulk
of others. The entire body is of very solid consistence, and of a pale reddish-
brown colour. The thorax or middle segment is armed with three pairs of
sharp spines; the head, also, has a pair of similar spines proceeding from
the cheeks behind. In our first walks we were puzzled to account for large
mounds of earth, of a different colour from the surrounding soil, which
were thrown up in the plantations and woods. Some of them were very
extensive, being forty yards in circumference, but not more than two feet
in height. We soon ascertained that these were the work of the Saübas,
being the outworks, or domes, which overlie and protect the entrances to
their vast subterranean galleries. On close examination, I found the earth of
which they are composed to consist of very minute granules, agglomerated
without cement, and forming many rows of little ridges and turrets. The
difference in colour from the superficial soil of the vicinity is owing to their
being formed of the undersoil, brought up from a considerable depth. It is
very rarely that the ants are seen at work on these mounds; the entrances
seem to be generally closed; only now and then, when some particular work
is going on, are the galleries opened. The entrances are small and numerous;
in the larger hillocks it would require a great amount of excavation to get
at the main galleries; but I succeeded in removing portions of the dome
in smaller hillocks, and then I found that the minor entrances converged,
at the depth of about two feet, to one broad elaborately worked gallery or
mine, which was four or five inches in diameter.

This habit in the Saüba ant of clipping and carrying away immense
quantities of leaves has long been recorded in books on natural history.
When employed on this work, their processions look like a multitude of
animated leaves on the march. In some places I found an accumulation of
such leaves, all circular pieces, about the size of a sixpence, lying on the

pathway, unattended by ants, and at some distance from any colony. Such heaps are always found to be removed when the place is revisited the next day. In course of time I had plenty of opportunities of seeing them at work. They mount the tree in multitudes, the individuals being all worker-minors. Each one places itself on the surface of a leaf, and cuts with its sharp scissor-like jaws a nearly semicircular incision on the upper side; it then takes the edge between its jaws, and by a sharp jerk detaches the piece. Sometimes they let the leaf drop to the ground, where a little heap accumulates, until carried off by another relay of workers; but, generally, each marches off with the piece it has operated upon, and as all take the same road to their colony, the path they follow becomes in a short time smooth and bare, looking like the impression of a cart-wheel through the herbage....

It has not hitherto been shown satisfactorily to what use it applies the leaves. I discovered it only after much time spent in investigation. The leaves are used to thatch the domes which cover the entrances to their subterranean dwellings, thereby protecting from the deluging rains the young broods in the nests beneath. The larger mounds, already described, are so extensive that few persons would attempt to remove them for the purpose of examining their interior; but smaller hillocks, covering other entrances to the same system of tunnels and chambers may be found in sheltered places, and these are always thatched with leaves, mingled with granules of earth. The heavily-laden workers, each carrying its segment of leaf vertically, the lower edge secured in its mandibles, troop up and cast their burdens on the hillock; another relay of labourers place the leaves in position, covering them with a layer of earthy granules, which are brought one by one from the soil beneath.

The underground abodes of this wonderful ant are known to be very extensive. The Rev. Hamlet Clark has related that the Saüba of Rio de Janeiro, a species closely allied to ours, has excavated a tunnel under the bed of the river Paráhyba, at a place where it is as broad as the Thames at London Bridge. At the Magoary rice mills, near Pará, these ants once pierced the embankment of a large reservoir: the great body of water which it contained escaped before the damage could be repaired. In the Botanic

Gardens, at Pará, an enterprising French gardener tried all he could think of to extirpate the Saüba. With this object he made fires over some of the main entrances to their colonies, and blew the fumes of sulphur down the galleries by means of bellows. I saw the smoke issue from a great number of outlets, one of which was seventy yards distant from the place where the bellows were used. This shows how extensively the underground galleries are ramified.

FOREST INSECTS

In these swampy shades [on the outskirts of Pará] we were afraid at each step of treading on some venomous reptile. On this first visit, however, we saw none, although I afterwards found serpents common here. We perceived no signs of the larger animals and saw very few birds. Insects were more numerous, especially butterflies. The most conspicuous species was a large, glossy, blue and black *Morpho* (*M. achilles*, of Linnaeus), which measures six inches or more in expanse of wings. It came along the alley at a rapid rate and with an undulating flight, but diverged into the thicket before reaching the spot where we stood. Another was the very handsome *Papilio sesostris*, velvety black in colour, with a large silky green patch on its wings. It is the male only which is so coloured; the female being plainer, and so utterly unlike its partner, that it was always held to be a different species until proved to be the same. Several other kinds allied to this inhabit almost exclusively these moist shades. In all of them the males are brilliantly coloured and widely different from the females. Such are *P. aeneas*, *P. vertumnus*, and *P. lysander*, all velvety black, with patches of green and crimson on their wings. The females of these species do not court the company of the males, but are found slowly flying in places where the shade is less dense. In the moist parts great numbers of males are seen, often four species together, threading the mazes of the forest, and occasionally rising to settle on the scarlet flowers of climbers near the tops of the trees. Occasionally a stray one is seen in the localities which the females frequent. In the swampiest parts, we saw numbers of the *Epicalia ancea*, one of the most richly coloured of the whole tribe of butterflies, being black, decorated

with broad stripes of pale blue and orange. It delighted to settle on the broad leaves of the Uraniae and similar plants where a ray of sunlight shone, but it was excessively wary, darting off with lightning speed when approached.

To obtain a fair notion of the number and variety of the animal tenants of these forests, it is necessary to follow up the research month after month and explore them in different directions and at all seasons. During several months I used to visit this district two or three days every week, and never failed to obtain some species new to me, of bird, reptile, or insect. It seemed to be an epitome of all that the humid portions of the Pará forests could produce. This endless diversity, the coolness of the air, the varied and strange forms of vegetation, the entire freedom from mosquitos and other pests, and even the solemn gloom and silence, combined to make my rambles through it always pleasant as well as profitable. Such places are paradises to a naturalist, and if he be of a contemplative turn there is no situation more favourable for his indulging the tendency. There is something in a tropical forest akin to the ocean in its effects on the mind. Man feels so completely his insignificance there, and the vastness of nature. A naturalist cannot help reflecting on the vegetable forces manifested on so grand a scale around him.

The Murderer Sipó

There is one kind of parasitic tree, very common near Pará, which exhibits this feature [fastening on others] in a very prominent manner. It is called the Sipó Matador, or the Murderer Liana. It belongs to the fig order, and has been described and figured by Von Martins in the Atlas to Spix and Martius's Travels. I observed many specimens. The base of its stem would be unable to bear the weight of the upper growth; it is obliged, therefore, to support itself on a tree of another species. In this it is not essentially different from other climbing trees and plants, but the way the matador sets about it is peculiar, and produces certainly a disagreeable impression. It springs up close to the tree on which it intends to fix itself, and the wood of its stem grows by spreading itself like a plastic mould over one side of the trunk of its supporter. It then puts forth, from each side, an arm-like

branch, which grows rapidly, and looks as though a stream of sap were flowing and hardening as it went. This adheres closely to the trunk of the victim and the two arms meet on the opposite side and blend together. These arms are put forth at somewhat regular intervals in mounting upwards, and the victim, when its strangler is full-grown, becomes tightly elapsed by a number of inflexible rings. These rings gradually grow larger as the Murderer flourishes, rearing its crown of foliage to the sky mingled with that of its neighbour, and in course of time they kill it by stopping the flow of its sap. The strange spectacle then remains of the selfish parasite clasping in its arms the lifeless and decaying body of its victim, which had been a help to its own growth. Its ends have been served — it has flowered and fruited, reproduced and disseminated its kind; and now, when the dead trunk moulders away, its own end approaches; its support is gone, and itself also falls. The Murderer Sipó merely exhibits, in a more conspicuous manner than usual, the struggle which necessarily exists amongst vegetable forms in these crowded forests, where individual is competing with individual and species with species, all striving to reach light and air in order to unfold their leaves and perfect their organs of fructification. All species entail in their successful struggles the injury or destruction of many of their neighbours or supporters, but the process is not in others so speaking to the eye as it is in the case of the Matador. The efforts to spread their roots are as strenuous in some plants and trees, as the struggle to mount upwards is in others. From these apparent strivings result the buttressed stems, the dangling air roots, and other similar phenomena. The competition amongst organised beings has been prominently brought forth in Darwin's *Origin of Species*; it is a fact which must be always kept in view in studying these subjects. It exists everywhere, in every zone, in both the animal and vegetable kingdoms.

Diurnal Cycle of Phenomena

We used to rise soon after dawn, when Isidoro would go down to the city, after supplying us with a cup of coffee, to purchase the fresh provisions for the day. The two hours before breakfast were devoted to ornithology. At that early period of the day the sky was invariably cloudless (the thermometer

marking 72° or 73° Fahr.); the heavy dew or the previous night's rain, which lay on the moist foliage, becoming quickly dissipated by the glowing sun, which rising straight out of the east, mounted rapidly towards the zenith. All nature was fresh, new leaf and flower-buds expanding rapidly. Some mornings a single tree would appear in flower amidst what was the preceding evening a uniform green mass of forest — a dome of blossom suddenly created as if by magic. The birds were all active; from the wild fruit trees, not far off, we often heard the shrill yelping of the Toucans (*Rhamphastos vitellinus*). Small flocks of parrots flew over on most mornings, at a great height, appearing in distinct relief against the blue sky, always two by two chattering to each other, the pairs being separated by regular intervals; their bright colours, however, were not apparent at that height. After breakfast we devoted the hours from 10 a.m. to 2 or 3 p.m. to entomology; the best time for insects in the forest being a little before the greatest heat of the day. We did not find them at all numerous, although of great variety as to species. The only kinds that appeared in great numbers were ants, termites, and certain species of social wasps; in the open grounds dragonflies were also amongst the most abundant kinds of insects. Beetles were certainly much lower in the proportion of individuals to species than they are in England, and this led us to the conclusion that the ants and termites here must perform many of the functions in nature which in temperate climates are the office of Coleoptera. As to butterflies, I extract the following note from many similar ones in my journal. "On Tuesday, collected 46 specimens, of 39 species. On Wednesday, 37 specimens, of 33 species, 27 of which are different from those taken on the preceding day." The number of specimens would be increased if I had reckoned all the commonest species seen, but still the fact is well established, that there is a great paucity of individuals compared with species in both Lepidoptera and Coleoptera.

FORMATION OF AMAZONS DELTA

The large collections which I made of the animal productions of Pará, especially of insects, enabled me to arrive at some conclusions regarding the relations of the Fauna of the south side of the Amazons Delta to those

of neighbouring regions. It is generally allowed that Guiana and Brazil, to the north and south of the Pará district, form two distinct provinces, as regards their animal and vegetable inhabitants. By this it is meant that the two regions have a very large number of forms peculiar to themselves, and which are supposed not to have been derived from other quarters during modern geological times. Each may be considered as a centre of distribution in the latest process of dissemination of species over the surface of tropical America. Pará lies midway between the two centres, each of which has a nucleus of elevated table-land, whilst the intermediate river-valley forms a wide extent of low-lying country. It is, therefore, interesting to ascertain from which the latter received its population, or whether it contains so large a number of endemic species as would warrant the conclusion that it is itself an independent province. To assist in deciding such questions as these, we must compare closely the species found in the district with those of the other contiguous regions, and endeavour to ascertain whether they are identical, or only slightly modified, or whether they are highly peculiar.

Von Martins, when he visited this part of Brazil forty years ago, coming from the south, was much struck with the dissimilarity of the animal and vegetable productions to those of other parts of Brazil. In fact, the Fauna of Pará, and the lower part of the Amazons, has no close relationship with that of Brazil proper; but it has a very great affinity with that of the coast region of Guiana, from Cayenne to Demerara. If we may judge from the results afforded by the study of certain families of insects, no peculiar Brazilian forms are found in the Pará district; whilst more than one-half the total number are essentially Guiana species, being found nowhere else but in Guiana and Amazonia. Many of them, however, are modified from the Guiana type, and about one-seventh seem to be restricted to Pará. These endemic species are not highly peculiar, and they may be yet found over a great part of northern Brazil when the country is better explored. They do not warrant us in concluding that the district forms an independent province, although they show that its Fauna is not wholly derivative, and that the land is probably not entirely a new formation. From all these facts, I think we must conclude that the Pará district belongs to the Guiana

province, and that, if it is newer land than Guiana, it must have received the great bulk of its animal population from that region. I am informed by Dr. Sclater that similar results are derivable from the comparison of the birds of these countries.

The interesting problem, how has the Amazons Delta been formed? receives light through this comparison of Faunas. Although the portion of Guiana in question is considerably nearer Pará than are the middle and southern parts of Brazil, yet it is separated from it by two wide expanses of water, which must serve as a barrier to migration in many cases. On the contrary, the land of Brazil proper is quite continuous from Rio Janeiro and Bahia up to Pará; and there are no signs of a barrier ever having existed between these places within recent geological epochs. Some of the species common to Pará and Guiana are not found higher up the river where it is narrower, so they could not have passed round in that direction. The question here arises, has the mouth of the Amazons always existed as a barrier to migration since the present species of the contiguous regions came into existence? It is difficult to decide the question; but the existing evidence goes far to show that it has not. If the mouth of the great river, which, for a long distance, is 170 miles broad, had been originally a wide gulf, and had become gradually filled up by islands formed of sediment brought down by the stream, we should have to decide that an effectual barrier had indeed existed. But the delta of the Amazons is not an alluvial formation like those of the Mississippi and the Nile. The islands in its midst and the margins of both shores have a foundation of rocks, which lie either bare or very near the surface of the soil. This is especially the case towards the sea-coast. In ascending the river southward and south-westward, a great extent of country is traversed which seems to have been made up wholly of river deposit, and here the land lies somewhat lower than it does on the sea-coast. The rocky and sandy country of Marajo and other islands of the delta towards the sea, is so similar in its physical configuration to the opposite mainland of Guiana that Von Martins concluded the whole might have been formerly connected, and that the Amazons had forced a way to the Atlantic through what was, perhaps, a close series of islands, or a continuous line of low country.

BIRD-KILLING SPIDER

In the course of our walk I chanced to verify a fact relating to the habits of a large hairy spider of the genus *Mygale*, in a manner worth recording. The species was *M. avicularia*, or one very closely allied to it; the individual was nearly two inches in length of body, but the legs expanded seven inches, and the entire body and legs were covered with coarse grey and reddish hairs. I was attracted by a movement of the monster on a tree-trunk; it was close beneath a deep crevice in the tree, across which was stretched a dense white web. The lower part of the web was broken, and two small birds, finches, were entangled in the pieces; they were about the size of the English siskin, and I judged the two to be male and female. One of them was quite dead, the other lay under the body of the spider not quite dead, and was smeared with the filthy liquor or saliva exuded by the monster. I drove away the spider and took the birds, but the second one soon died. The fact of species of *Mygale* sallying forth at night, mounting trees, and sucking the eggs and young of humming-birds, has been recorded long ago by Madame Merian and Palisot de Beauvois; but, in the absence of any confirmation, it has come to be discredited. From the way the fact has been related it would appear

Bird-killing spider attacking finches

that it had been merely derived from the report of natives, and had not been witnessed by the narrators. Count Langsdorff, in his *Expedition into the Interior of Brazil*, states that he totally disbelieved the story. I found the circumstance to be quite a novelty to the residents hereabout. The *Mygales* are quite common: some species make their cells under stones, others form artistical tunnels in the earth, and some build their dens in the thatch of houses. The natives call them Aranhas carangueijeiras, or crab-spiders. The hairs with which they are clothed come off when touched, and cause a peculiar and almost maddening irritation. The first specimen that I killed and prepared was handled incautiously, and I suffered terribly for three days afterwards. I think this is not owing to any poisonous quality residing in the hairs, but to their being short and hard, and thus getting into the fine creases of the skin. Some *Mygales* are of immense size. One day I saw the children belonging to an Indian family who collected for me with one of these monsters secured by a cord round its waist, by which they were leading it about the house as they would a dog.

Humming-Birds

In January the orange-trees became covered with blossom — at least to a greater extent than usual, for they flower more or less in this country all the year round — and the flowers attracted a great number of humming-birds. Every day, in the cooler hours of the morning, and in the evening from four o'clock till six, they were to be seen whirring about the trees by scores. Their motions are unlike those of all other birds. They dart to and fro so swiftly that the eye can scarcely follow them, and when they stop before a flower it is only for a few moments. They poise themselves in an unsteady manner, their wings moving with inconceivable rapidity, probe the flower, and then shoot off to another part of the tree. They do not proceed in that methodical manner which bees follow, taking the flowers seriatim, but skip about from one part of the tree to another in the most capricious way. Sometimes two males close with each other and fight, mounting upwards in the struggle as insects are often seen to do when similarly engaged, and then separating hastily and darting back to their work. Now and then they

stop to rest, perching on leafless twigs, when they may be sometimes seen probing, from the place where they sit, the flowers within their reach. The brilliant colours with which they are adorned cannot be seen whilst they are fluttering about, nor can the different species be distinguished unless they have a deal of white hue in their plumage, such as *Heliothrix auritus*, which is wholly white underneath although of a glittering green colour above, and the white-tailed *Florisuga mellivora*. There is not a great variety of humming-birds in the Amazons region, the number of species being far smaller in these uniform forest plains than in the diversified valleys of the Andes, under the same parallels of latitude. The family is divisible into two groups contrasted in form and habits, one containing species which live entirely in the shade of the forest, and the other comprising those which prefer open sunny places. The forest species (*Phaethominae*) are seldom seen at flowers, flowers being, in the shady places where they abide, of rare occurrence; but they search for insects on leaves, threading the bushes and passing above and beneath each leaf with wonderful rapidity. The other group (*Trochilinae*) are not quite confined to cleared places, as they come into the forest wherever a tree is in blossom, and descend into sunny openings where flowers are to be found. But it is only where the woods are less dense than usual that this is the case; in the lofty forests and twilight shades of the low lands and islands they are scarcely ever seen. I searched well at Caripí, expecting to find the *Lophornis gouldii*, which I was told had been obtained in the locality. This is one of the most beautiful of all humming-birds, having round its neck a frill of long white feathers tipped with golden green. I was not, however, so fortunate as to meet with it. Several times I shot by mistake a humming-bird hawk-moth instead of a bird. This moth (*Macroglossa titan*) is somewhat smaller than humming-birds generally are, but its manner of flight, and the way it poises itself before a flower whilst probing it with its proboscis are precisely like the same actions of humming-birds. It was only after many days' experience that I learnt to distinguish one from the other when on the wing. This resemblance has attracted the notice of the natives, all of whom, even educated whites, firmly believe that one is transmutable into the other. They have observed the metamorphosis

Humming-bird (right) and Humming-bird Hawk-moth

of caterpillars into butterflies, and think it not at all more wonderful that a moth should change into a humming-bird. The resemblance between this hawk-moth and a humming-bird is certainly very curious, and strikes one even when both are examined in the hand. Holding them sideways, the shape of the head and position of the eyes in the moth are seen to be nearly the same as in the bird, the extended proboscis representing the long beak. At the tip of the moth's body there is a brush of long hair-scales resembling feathers, which, being expanded, looks very much like a bird's tail. But, of course, all these points of resemblance are merely superficial.

Beautiful Vegetation

After dinner I expressed a wish to see more of the creek, so a lively and polite old man, whom I took to be one of the neighbours, volunteered as guide. We embarked in a little montaria, and paddled some three or four miles up and down the stream. Although I had now become familiarised with beautiful vegetation, all the glow of fresh admiration came again to me in this place.

The creek was about 100 yards wide, but narrower in some places. Both banks were masked by lofty walls of green drapery, here and there a break occurring through which, under over-arching trees, glimpses were obtained of the palm-thatched huts of settlers. The projecting boughs of lofty trees, which in some places stretched half-way across the creek, were hung with natural garlands and festoons, and an endless variety of creeping plants clothed the water frontage, some of which, especially the *Bignonias*, were ornamented with large gaily-coloured flowers. Art could not have assorted together beautiful vegetable forms so harmoniously as was here done by Nature. Palms, as usual, formed a large proportion of the lower trees; some of them, however, shot up their slim stems to a height of sixty feet or more, and waved their bunches of nodding plumes between us and the sky. One kind of palm, the Pashiúba (*Iriartea exorrhiza*), which grows here in greater abundance than elsewhere, was especially attractive. It is not one of the tallest kinds, for when full-grown its height is not more, perhaps, than forty feet; the leaves are somewhat less drooping, and the leaflets much broader than in other species, so that they have not that feathery appearance which those of some palms have, but still they possess their own peculiar beauty.

COLOURS OF ANIMALS

In the sandy beach I found two species of *Tetracha*, a genus of tiger-beetles, which have remarkably large heads, and are found only in hot climates. They come forth at night, in the daytime remaining hid in their burrows several inches deep in the light soil. Their powers of running exceed everything I witnessed in this style of insect locomotion. They run in a serpentine course over the smooth sand, and when closely pursued by the fingers in the endeavour to seize them, are apt to turn suddenly back, and thus baffle the most practised hand and eye. I afterwards became much interested in these insects on several accounts, one of which was that they afforded an illustration of a curious problem in natural history. One of the Caripí species (*T. nocturna* of Dejean) was of a pallid hue like the sand over which it ran; the other was a brilliant copper-coloured kind (*T. pallipes* of Klug). Many insects whose abode is the sandy beaches are white in colour; I

found a large earwig and a mole-cricket of this hue very common in these localities. Now it has been often said, when insects, lizards, snakes, and other animals, are coloured so as to resemble the objects on which they live, that such is a provision of nature, the assimilation of colours being given in order to conceal the creatures from the keen eyes of insectivorous birds and other animals. This is no doubt the right view, but some authors have found a difficulty in the explanation on account of this assimilation of colours being exhibited by some kinds and not by others living in company with them; the dress of some species being in striking contrast to the colours of their dwelling-place. One of our *Tetracha* is coloured to resemble the sand, whilst its sister species is a conspicuous object on the sand; the white species, it may be mentioned, being much more swift of foot than the copper-coloured one. The margins of these sandy beaches are frequented throughout the fine season by flocks of sandpipers, who search for insects on moonlit nights as well as by day. If one species of insect obtains immunity from their onslaughts by its deceptive resemblance to the sandy surface on which it runs, why is not its sister species endowed in the same way? The answer is, that the dark-coloured kind has means of protection of quite a different nature, and therefore does not need the peculiar mode of disguise enjoyed by its companion. When handled it emits a strong, offensive, putrid and musky odour, a property which the pale kind does not exhibit. Thus we see that the fact of some species not exhibiting the same adaptation of colours to dwelling-places as their companion species does not throw doubt on the explanation given of the adaptation, but is rather confirmatory of it.

Musical Cricket

A strange kind of wood-cricket is found in this neighbourhood. The males produce a very loud and not unmusical noise by rubbing together the overlapping edges of their wing-cases. The notes are certainly the loudest and most extraordinary that I ever heard produced by an orthopterous insect. The natives call it the Tananá, in allusion to its music, which is a sharp, resonant stridulation resembling the syllables ta-na-ná, ta-na-ná, succeeding each other with little intermission. It seems to be rare in the

neighbourhood. When the natives capture one they keep it in a wicker-work cage for the sake of hearing it sing. A friend of mine kept one six days. It was lively only for two or three, and then its loud note could be heard from one end of the village to the other. When it died he gave me the specimen, the only one I was able to procure. It is a member of the family Locustidae, a group intermediate between the Crickets (Achetidae) and the Grasshoppers (Acridiidae). The total length of the body is two inches and a quarter; when the wings are closed the insect has an inflated vesicular or bladder-like shape, owing to the great convexity of the thin but firm parchmenty wing-cases, and the colour is wholly pale-green. The instrument by which the Tananá produces its music is curiously contrived out of the ordinary nervures of the wing-cases. In each wing-case the inner edge, near its origin, has a horny expansion or lobe; on one wing this lobe has sharp raised margins; on the other, the strong nervure which traverses the lobe on the underside is crossed by a number of fine sharp furrows like those of a file. When the insect rapidly moves its wings, the file of the one lobe is scraped sharply across the horny margin of the other, thus producing the sounds; the parchmenty wing-cases and the hollow drum-like space which they enclose assisting to give resonance to the tones. The projecting portions of both wing-cases are traversed by a similar strong nervure, but this is scored like a file in one of them, in the other remaining perfectly smooth.

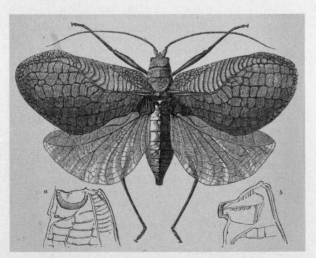

Musical cricket
(*Chlorocoelus tananá*)
a. b. Lobes of wing-cases
transformed into a musical
instrument

ORIGIN OF SPECIES

The way in which modifications [of varieties of species] occur [in Obydos] merits a few remarks. I will therefore give an account of one very instructive case which presented itself in this neighbourhood.

The case was furnished by certain kinds of handsome butterflies belonging to the genus *Heliconius*, a group peculiar to Tropical America, abounding in individuals everywhere in the shades of its luxuriant forests, and presenting clusters of varieties and closely allied species, as well as many distinct, better marked forms. The closely allied species and varieties are a great puzzle to classifiers; in fact, the group is one of those wherein great changes seem to be now going on. A conspicuous member of the group is the *H. melpomene* of Linnaeus. This elegant form is found throughout Guiana, Venezuela, and some parts of New Granada. It is very common at Obydos, and reappears on the south side of the river in the dry forests behind Santarém, at the mouth of the Tapajós. In all other parts of the Amazons valley, eastward to Pará and westward to Peru, it is entirely absent. This absence at first appeared to me very strange; for the local conditions of these regions did not appear so strongly contrasted as to check, in this abrupt manner, the range of so prolific a species; especially as at Obydos and Santarém it occurred in moist woods close to the edge of the river. Another and nearly allied species, however, takes its place in the forest plains; namely, the *H. thelxiope* of Hübner. It is of the same size and shape as its sister kind, but differs very strikingly in colours: *H. melpomene* being simply black with a large crimson spot on its wings, whilst *H. thelxiope* has these beautifully rayed with black and crimson, and is further adorned with a number of bright yellow spots. Both have the same habits. *H. melpomene* ornaments the sandy alleys in the forests of Obydos, floating lazily in great numbers over the lower trees; whilst *H. thelxiope*, in a similar manner and in equal numbers, adorns the moister forests which constitute its domain. No one who has studied the group has doubted for a moment that the two are perfectly and originally distinct species, like the hare and rabbit, for instance, or any other two allied species of one and the same genus. The following facts, however, led me to conclude that the one is simply a modification of the other. There are, as

Heliconius thelxiope   *Heliconius melpomene*

might be supposed, districts of forest intermediate in character between the drier areas of Obydos, &c., and the moister tracts which compose the rest of the immense river valley. At two places in these intermediate districts, namely, Serpa, 180 miles west of Obydos, on the same side of the river, and Aveyros, on the lower Tapajós, most of the individuals of these *Heliconii* which occurred were transition forms between the two species. Already, at Obydos, *H. melpomene* showed some slight variation amongst its individuals in the direction of *H. thelxiope*, but not anything nearly approaching it. It might be said that these transition forms were hybrids, produced by the intercrossing of two originally distinct species; but the two come in contact in several places where these intermediate examples are unknown, and I never observed them to pair with each other. Besides which, many of them occur also on the coast of Guiana, where *H. thelxiope* has never been found. These hybrid-looking specimens are connected together by so complete a chain of gradations that it is difficult to separate them even into varieties, and they are incomparably more rare than the two extreme forms. They link together gradually the wide interval between the two species. One is driven to conclude that the two were originally one and the same: the mode in which they occur and their relative geographical positions being in favour of the supposition that *H. thelxiope* has been derived from *H. melpomene*. Both are nevertheless good and true species in all the essential characters of species; for, as already observed, they do not pair together when existing side by side, nor is there any appearance of reversion to an original common form under the same circumstances.

In the controversy which is being waged amongst Naturalists, since the publication of the Darwinian theory of the origin of species, it has

been rightly said that no proof at present existed of the production of a physiological species, — that is, a form which will not interbreed with the one from which it was derived, although given ample opportunities of doing so, and does not exhibit signs of reverting to its parent form when placed under the same conditions with it. Morphological species, — that is, forms which differ to an amount that would justify their being considered good species, have been produced in plenty through selection by man out of variations arising under domestication or cultivation. The facts just given are, therefore, of some scientific importance; for they tend to show that a physiological species can be and is produced in nature out of the varieties of a pre-existing closely allied one. This is not an isolated case; for I observed, in the course of my travels, a number of similar instances. But in very few has it happened that the species which clearly appears to be the parent coexists with one that has been evidently derived from it. Generally the supposed parent also seems to have been modified, and then the demonstration is not so clear, for some of the links in the chain of variation are wanting. The process of origination of a species in nature, as it takes place successively, must be ever perhaps beyond man's power to trace, on account of the great lapse of time it requires. But we can obtain a fair view of it by tracing a variable and far-spreading species over the wide area of its present distribution; and a long observation of such will lead to the conclusion that new species in all cases must have arisen out of variable and widely disseminated forms. It sometimes happens, as in the present instance, that we find in one locality a species under a certain form which is constant to all the individuals concerned; in another exhibiting numerous varieties; and in a third presenting itself as a constant form, quite distinct from the one we set out with. If we meet with any two of these modifications living side by side, and maintaining their distinctive characters under such circumstances, the proof of the natural origination of a species is complete: it could not be much more so were we able to watch the process step by step. It might be objected that the difference between our two species is but slight, and that by classing them as varieties nothing further would be proved by them. But the differences between them are such as obtain between allied species

generally. Large genera are composed, in great part, of such species; and it is interesting to show how the great and beautiful diversity within a large genus is brought about by the working of laws within our comprehension.

TICKS

The higher and drier land is everywhere sandy, and tall coarse grasses line the borders of the broad alleys which have been cut through the second-growth woods. These places swarm with carrapatos, ugly ticks belonging to the genus *Ixodes*, which mount to the tips of blades of grass, and attach themselves to the clothes of passers by. They are a great annoyance. It occupied me a full hour daily to pick them off my flesh after my diurnal ramble. There are two species; both are much flattened in shape, have four pairs of legs, a thick short proboscis and a horny integument. Their habit is to attach themselves to the skin by plunging their proboscides into it, and then suck the blood until their flat bodies are distended into a globular form. The whole proceeding, however, is very slow, and it takes them several days to pump their fill. No pain or itching is felt, but serious sores are caused if care is not taken in removing them, as the proboscis is liable to break off and remain in the wound. A little tobacco juice is generally applied to make them loosen their hold. They do not cling firmly to the skin by their legs, although each of these has a pair of sharp and fine claws connected with the tips of the member by means of a flexible pedicle. When they mount to the summits of slender blades of grass, or the tips of leaves, they hold on by their fore legs only, the other three pairs being stretched out so as to fasten on any animal which comes in their way. The smaller of the two species is of a yellowish colour; it is much the most abundant, and sometimes falls upon one by scores. When distended it is about the size of a No. 8 shot; the larger kind, which fortunately comes only singly to the work, swells to the size of a pea.

MIMETIC RESEMBLANCES

One of the most conspicuous insects peculiar to Villa Nova is an exceedingly handsome butterfly, which has been named *Agrias phalcidon*. It is of large size, and the colours of the upper surface of its wings resemble those of

the *Callithea leprieurii*, already described, namely, dark blue, with a broad silvery-green border. When it settles on leaves of trees, fifteen or twenty feet from the ground, it closes its wings and then exhibits a row of brilliant pale-blue eye-like spots with white pupils, which adorns their under surface. Its flight is exceedingly swift, but when at rest it is not easily made to budge from its place; or if driven off, returns soon after to the same spot. Its superficial resemblance to *Callithea leprieurii*, which is a very abundant species in the same locality, is very close. The likeness might be considered a mere accidental coincidence, especially as it refers chiefly to the upper surface of the wings, if similar parallel resemblances did not occur between other species of the same two genera. Thus, on the Upper Amazons, another totally distinct kind of *Agrias* mimicks still more closely another *Callithea*; both insects being peculiar to the district where they are found flying together. Resemblances of this nature are very numerous in the insect world. I was much struck with them in the course of my travels, especially when, on removing from one district to another, local varieties of certain species were found accompanied by local varieties of the species which counterfeited them in the former locality, under a dress changed to correspond with the altered liveries of the species they mimicked. One cannot help concluding these imitations to be intentional, and that nature has some motive in their production. In many cases, the reason of the imitation is sufficiently plain. For instance, when a fly or psitic bee has a deceptive resemblance to the species of working bee, in whose nest it deposits the egg it has otherwise no means of providing for, or when a leaping-spider, as it crouches in the axil of a leaf waiting for its prey, presents an exact imitation of a flower-bud; it is evident that the benefit of the imitating species is the object had in view. When, however, an insect mimics another species of its own order where predaceous or parasitic habits are out of the question, it is not so easy to divine the precise motive of the adaptation. We may be sure, nevertheless, that one of the two is assimilated in external appearance to the other for some purpose useful, — perhaps of life and death importance — to the species. I believe these imitations are of the same nature as those in which an insect or lizard is coloured and marked so as to resemble the

soil, leaf, or bark on which it lives; the resemblance serving to conceal the creatures from the prying eyes of their enemies; or, if they are predaceous species, serving them as a disguise to enable them to approach their prey. When an insect, instead of a dead or inorganic substance, mimics another species of its own order, and does not prey, or is not parasitic, may it not be inferred that the mimicker is subject to a persecution by insectivorous animals from which its model is free? Many species of insects have a most deceptive resemblance to living or dead leaves; it is generally admitted, that this serves to protect them from the onslaughts of insect-feeding animals who would devour the insect, but refuse the leaf. The same might be said of a species mimicking another of the same order; one may be as repugnant to the tastes of insect persecutors, as a leaf or a piece of bark would be, and its imitator not enjoying this advantage would escape by being deceptively assimilated to it in external appearances. In the present instance, it is not very clear what property the *Callithea* possesses to render it less liable to persecution than the *Agrias*, except it be that it has a strong odour somewhat resembling Vanilla, which the *Agrias* is destitute of. This odour becomes very powerful when the insect is roughly handled or pinched, and if it serves as a protection to the *Callithea*, it would explain why the *Agrias* is assimilated to it in colours. The resemblance, as before remarked, applies chiefly to the upper side; in other species it is equally close on both surfaces of the wings. Some birds, and the great Aeschnae dragon-flies, take their insect prey whilst on the wing, when the upper surface of the wings is the side most conspicuous.

Piúm Flies

We made acquaintance on this coast with a new insect pest, the Piúm, a minute fly, two-thirds of a line in length, which here commences its reign, and continues henceforward as a terrible scourge along the upper river, or Solimoens, to the end of the navigation on the Amazons. It comes forth only by day, relieving the mosquito at sunrise with the greatest punctuality, and occurs only near the muddy shores of the stream, not one ever being found in the shade of the forest. In places where it is abundant it accompanies

canoes in such dense swarms as to resemble thin clouds of smoke. It made its appearance in this way the first day after we crossed the river. Before I was aware of the presence of flies, I felt a slight itching on my neck, wrist, and ankles, and on looking for the cause, saw a number of tiny objects having a disgusting resemblance to lice, adhering to the skin. This was my introduction to the much-talked of Piúm. On close examination, they are seen to be minute two-winged insects, with dark-coloured body and pale legs and wings, the latter closed lengthwise over the back. They alight imperceptibly, and squatting close, fall at once to work; stretching forward their long front legs, which are in constant motion and seem to act as feelers, and then applying their short, broad snouts to the skin. Their abdomens soon become distended and red with blood, and then, their thirst satisfied, they slowly move off, sometimes so stupefied with their potations that they can scarcely fly. No pain is felt whilst they are at work, but they each leave a small circular raised spot on the skin and a disagreeable irritation. The latter may be avoided in great measure by pressing out the blood which remains in the spot; but this is a troublesome task when one has several hundred punctures in the course of a day. I took the trouble to dissect specimens to ascertain the way in which the little pests operate. The mouth consists of a pair of thick fleshy lips, and two triangular horny lancets, answering to the upper lip and tongue of other insects. This is applied closely to the skin, a puncture is made with the lancets, and the blood then sucked through between these into the oesophagus, the circular spot which results coinciding with the shape of the lips.

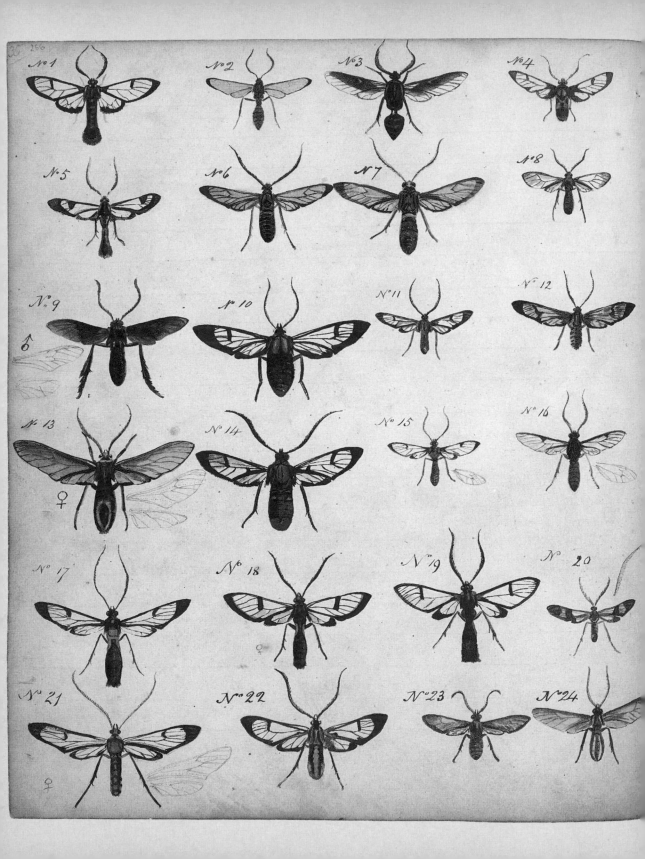

Insect Fauna
of
the Amazon Valley

A SELECTION OF PAGES FROM THE FIRST JOURNAL

Gymnetis

Melolonthidae

Semiotus — The clypeus projects over
front in form of two teeth: between it &
base of labrum is the perpendicular
epistome. The labrum is as broad as
long, rounded in front, attached close
under it is a membrane bifid on
its front edge. Mandibles are very
broad at base, suddenly very much
narrowed in middle & terminated
in a bifid point. The maxillae
are feeble — the blades thin & almost
entirely composed of a flattened pencil
of hairs. The labium consists — 1° basal
of mentum flexible & horny — 2° the tumid
membranous
labrum & arising from top edge of mentum
& terminating in the two supports of palpi
3° the ligula, arising from the mouth
within & attached to inner side of
labrum (apparently to me it springs from
root of mentum): it is a deeply
cleft thin transparent membrane
the upper edge of the lobes fringed with
hairs.

Legs. — The trochanters are remarkably
long, causing the femora to be distant &
removed from coxa. The 3 basal jts.
of tarsi are flapped.

The prosternum is produced into a
chin cover — The projection is as
in sketch. The cotyloid cavities
are open behind.

The sides of base of elytra lapp
over the epimera sides of metasternum
— The species is a common one
Oct 4 / 55

labrum

mandible

Prothx.

max

lab. & ment
caticula

Gymnetis (pale-greenish-ashy-gray
many at Ega). —
mandible very curious, a narrow
thistle-horny blade, with a soft membranous expansion produced from the
inner side, the base
common to these 2 parts being hard
& horny: the whole concealed under
the clypeus. labrum is also
wholly soft & membranous. The
labial structure as usual, merits
the most attention: the structure it-
being that of Lamellicornes
generally. The whole organ
appears to be one horny hard
plate, the ligula (or paraglossa, being
bipartite) & the soft basal parts of
ligula from which spring the palpi,
& the whole of which in most
Coleoptera (or Longicornes passim) are
extended & exposed beyond tip of
mentum, all being here soldered
to the innerside of the horny plate or organs of stridut.
mentum, the apices of ligula
(or paraglossa) only being free. The
palpi escape by a cleft left on
each side near the tip of the mentum
& between it & the paraglossa.
April 29 1856

Isonychus? The mandibles
offer 3 parts like those of Gymnetis
— 1 An outer narrow horny piece
which in this genus is trigonal
2 a broad membranous expansion
arising from
attached the inner side of 1, &
which in this case is equal in
height to it — 3 a hard thick
horny piece serving as a common
base to 1 & 2. — The maxillae
want the outer lobe, the inner
is extremely short & furnished
with 5 or 6 long, acute, teeth.
The mentum appears to be round
off at its upper angles & the
upper edge of labium (which is
here uncleft) is rather higher th
it. — Ega 12 May 1856

Ligyrus? (Pentodontides) — no
Mand. visible on
clyp. Ega 3 August /56

Phaneus

maxilla

palp
2 lobe of lig
— integument
of throat

corect

mentum

produced   groove   epistome
            antenn   + labrum

Clivina

lab. palps. 225   4

No 1

mendibles   ligula, central
            part more corneous
            p.p. paraglossae, very thin
            white membranous, distinct
            from base
            lab. palps. termi. jt. enlarged
            concave on outer face

Phaneus — (sloping plane prons.)
1 in the mentum sc. Outside face
the palpi arise from the thick
& inflated membranous space
situated the inner face of mentum
& the ligula (see 3 side views) but
the whole of the 1st jt. of palps, is
apparent above the ment. the
ligula is attached inside of ment.
& inflated membranous space, root
of lig. coincident with root or
even hinge of mentum. lig. is broadly
divided into 2 narrow lobes
membranous, & ciliated as is usual
having one narrow horny edge
(2) — maxillae have the stem very
thick & horny — the lobes soft
membranous, inner lobe nearly
ovular forming on its inner face
ie. that which faces the labrum
a flat cushion of soft hairs or
capilla. Ega 2 June 1856

Lab. Mentum
& Labipalps of
Eurysternus, shewing the
deficiency of terminal jt. of 3rd
palps — & the length & narrowness of the lobes of ligula
behind, which are longer
than the palpi
June 3 1856

Scarítida — found at Ega — attracted
to lamp at night in July & March — &
thrown up on sandy beach by storm at
night in August — the Maxilla are
peculiarly small & slender & wholly within side
mentum. the ligula concealed behind
mentum — in 3 dissection could not make into
the shape   May 21/57

species of small Scarites
size of Clivina. broad flat
prothorax — body flat
all black   fore tibia
only 2 distinct teeth outside
Ega 23 August 1856 — spoilt
mentum in dissection
but observed the ligula
distinctly

The ligula is exsertile & contractile
have now dissected 5 specns. Think
the diff in lab. palpi is sexual
Aug. 2/56

epistome
& labrum   maxilla

fore leg   Nat. size
            shining
            ferruginous

No 2

mentum
& labi.palps of other individ.
of same species

Clivina?   In dissecting
insects having the facies of
European Clivina & Dyschirius
I do not find the parts
agree with any of the descriptions
of Putzeys. — The present
specs (common at Ega) is
exactly like a Clivina in facies,
depressed protho. ferruginous
colour: But the Labial palpi
differ in the individs. about ½
of them havg. termi. jt. much
dilated & concave in exterior
face, other ½ have it oblong
nearly cylindric. The other
parts are as I have attached
them. The ligula is much
drawn down in drawing of mentum
in No 2 than No 1 But shape
the same — Ega 30 July 1856

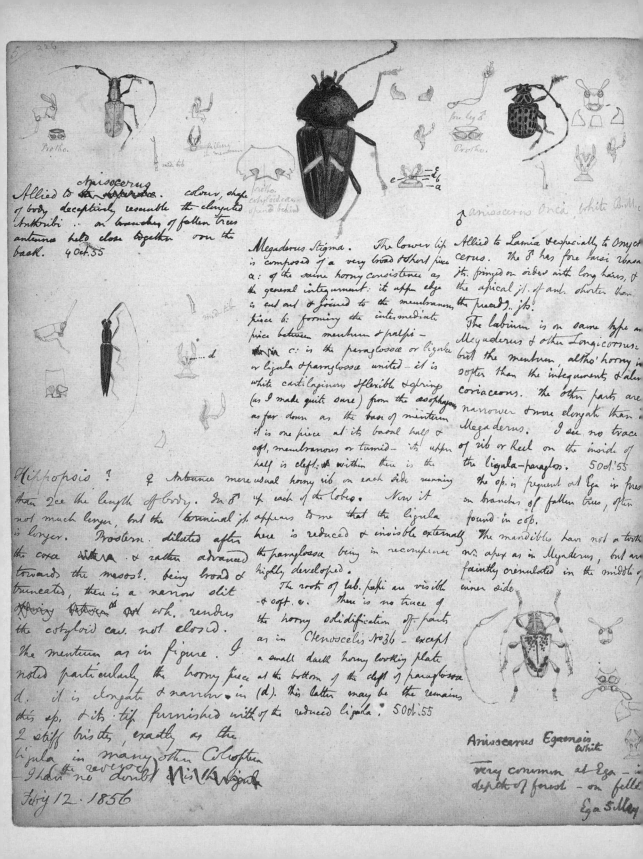

5. 226

Anisocerus

Allied to ~~the species~~. colour, shape of body deceptively resemble the elongated *Onithribi* :- on branches of fallen trees antennae held close together over the back. 4 Oct. 55

Protho.
mid. tib.
...palpi ...mentum
Protho. cotyloidcav open behind

Megaderus Stigma. The lower lip is composed of a very broad & short piece a: of the same horny consistence as the general integument: its upper edge is cut out & joined to the membranous piece b: forming the intermediate piece between mentum & palpi —
~~on~~ c: is the paraglossae or ligula or ligula & paraglossae united — it is white cartilaginous flexible & spring (as I made quite sure) from the œsophagus as far down as the base of mentum it is one piece at its basal half & soft, membranous or tumid — its upper half is cleft: & within there is the more usual horny rib on each side running up each of the lobes. Now it appears to me that the ligula here is reduced & invisible externally the paraglossae being in recompense highly developed.
The roots of lab. palpi are visible & soft. e. There is no trace of the horny solidification of parts as in *Ctenoscelis* No 36 — except a small darkish horny looking plate at the bottom of the cleft of paraglossa in (d). This latter may be the remains of the reduced ligula. 5 Oct. 55

♂ *Anisocerus Onca.* White B. M.

Allied to *Lamia* & especially to *Onychocerus.* The ♂ has fore tarsi ~~basal~~ its fringed on sides with long hairs, & the apical jt. of ant. shorter than the preced. jt.
The labrium is on same type as *Megaderus* & other Longicorns: but the mentum, altho' horny is softer than the integuments & also coriaceous. The other parts are narrower & more elongate than in *Megaderus.* I see no trace of rib or keel on the inside of the ligula-paraglossa. 5 Oct. 55
The sp. is frequent at Ega in ~~fores~~ on branches of fallen trees, often found in cop.
The mandibles have not a tooth nr. apex as in *Megaderus*, but are faintly crenulated in the middle of inner side.

Hippopsis? ♀ Antennae more than 2ce the length of body. In ♂ not much longer, but the terminal jt. is longer. Prosternum dilated after the coxae ~~at the~~ & rather advanced towards the mesost. being broad & truncated, there is a narrow slit ~~dividing between~~ wh. renders the cotyloid cav. not closed.
The mentum as in figure. I noted particularly the horny piece d. It is elongate & narrow in this sp. & its tip furnished with 2 stiff bristles, exactly as the ligula in many other Coleoptera — I have no doubt ~~the reverse~~ ...

Feby 12. 1856

Anisocerus Egaensis White

very common at Ega — in depth of forest — on felle
Ega 5 May

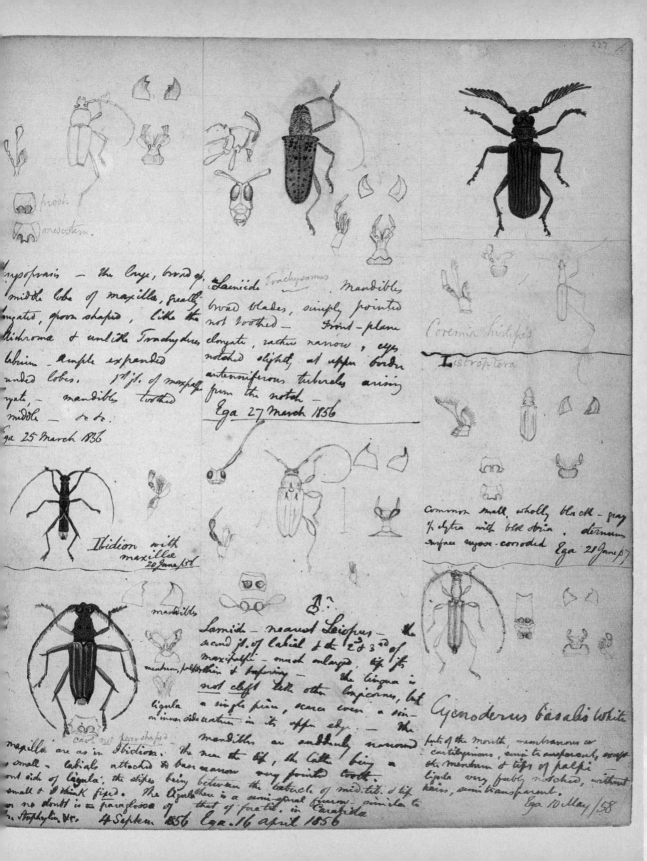

frost.

mesostern.

Hypoprais — the large, broad of, middle lobe of maxilla, greatly elongated, spoon shaped, like the Michroma & unlike Trachyderes labium — Ample expanded rounded lobes. 1st js. of maxpalp ..yate, — mandibles toothed middle — &c &c.
Ega 25th March 1856

Laeniide Trachysomus. Mandibles broad blades, simply pointed not toothed — Front-plane elongate, rather narrow, eyes notched slightly, at upper border antenniferous tubercles arising from the notch —
Ega 27 March 1856

Coremia hirtipes

Distroptera

common small, wholly black - gray ½ elytra with black stria. sternum surface rugose-corroded Ega 21 June 57

Ibidion with maxillæ 20 June 56

mandibles

Lamiide — nearest Leiopus — the second js. of labial & the 2d & 3d of maxipalpi — much enlarged, tip js. mentum, palpi thin & tapering — the lingua is not cleft like other longicorns, but a single piece, scarce even a sin- ligula on inner side & rather in its upper edge. — The on inner side mandibles are suddenly narrowed maxilla are as in Ibidion; the near the tip, the latter being a .. small - labials attached to base narrow very pointed tooth. .. out side of ligula; the stipes being between the tentacle of mid.tib. & tip small & I think fixed. The ligula there is a semi-spiral tinsion- similar to .. no doubt is the paraglossa of that of fore tib. in Carabida .. Staphylini &c. 4 Septem 1856 Ega 16 april 1856

cav, not pear-shaped

Cyenoderus basalis White parts of the mouth, membranous or cartilaginous, semi transparent, except the mentum & tips of palpi ligula very faintly notched, without hairs, semi transparent.
Ega 10 May/58

antenna

Ctenostoma formicarium
or *Procephalus sp.?*
Decidedly destitute of the
maxillary claw. I suppose
this is ♂, the 2–4 jts of fore
tarsi are covered with not
dense. palpi, 2 jt. thick, 3rd
very short felt beneath. The
3.4 & jt of other tarsi have
also a short pile beneath.
the 4th is slightly sinuated.
The blade of the maxilla
is blunt, thick & only slightly
curved, (to Carabici it is generally
thin, acute & much curved)
the 2nd jts both of max. &
labi.palps, have a double
two
divergent series of strong
spinous hairs.
We find, therefore, the hook of
max. less a distinction of
Cicindelida than the invisibility
of labial root & paraglossæ —
deeply excavate mentum —
want of notch to fore tibia —
& toothed mandible. —
May 4. 1853

Dromius, one of the bark species
— mouth much flattened both
above & beneath. — Mandibles
thin blades, broad & thin at the
base, 1/3rd covered by the labrum
— Maxilla with the jt. projecting
onward, in an acute elbow
— basal jt narrow, blade narrow
rather straight, tip fine
curved, acute, ciliæ short, not
dense. palpi, 2 jt. thick, 3rd &
4th forming together an elongate
oval, point truncate, of which
3rd jt. forms 1/3 & terml. jt. 2/3rd
of the length. Mentum small
sides rounded, sinuation shallow
semicircular (? a minute tooth
generally in middle) — Labium elongate
4 Dente, ligula narrow, paraglossæ
thin, membranous, longer than
ligula, appearing glabrous.

N.B. the structure of the
mouth, evidently adapted to
its mode of taking its prey
— under the bark of trees.
July 9. 1853

labium
labium & paraglossæ

labium & paraglossæ
of another species

These mouths appear to agree with the
defin. of Dromius in everything except
the labrum which here is always elongate
covering nearly whole of mandible, narrowed
forward & unevenly cut in front. The
languette is certainly much enveloped
by paraglossæ. From these characters
they are Perricallidæ. but the mentum
of 4 species examined has no tooth
in centre.
27 Octob. 1853

prothox. adp. opened (possibly Lophocerus bicornis
Allied to Trachyderes — much covered
with golden-fulvous pubescence. especially
antennæ tibiæ & tarsi, the antennæ
have a longer pubescence beneath their
joints & the tarsi cushion of shorter
pile beneath. The terminal joints
are enlarged from base to apex & sharply
truncate, compressed. The lobe of max.
is broad, truncate, very little longer than
the blade, then differing greatly from
true Cerambycine group.
The differences from true Trachyderes
(T. succinctus) are (1) mandibles simply
pointed not bifid. (2) mentum sides
rounded not angular (3) scutellum very
(tarsi. jt cylindric — beneath not compress
or sinuate — & &c.
Day-flying insect like the Trachyde
generally — about fallen trees &c. in for
Septem 24/1853

made a better dissection — The acetylin
of prothox. open behind like Megaderma
Parts of mouth all small. mentum is
very broadly excavated on upp. edge like Megad
from this arise the turned membranous
supporting the labial appendages. there is
central piece in front at base of the clype
paraglossa & between the root of palpi, which I
in other Longicornes. it is horny. Here it is
apparent because pubescent like the other
parts of their organs, the membranous parts
glabrous.   Oct. 10 1855

Lophonocerus barbico.
made here a very satisfactory dissection.
piece in front between two palpi &at base
cleft of paraglossæ, I see is notched
Oct 10 – 55

Pæcilopeplus Batesii ♂

Mandibles short triangular, the
sinistral pointed, the dextral
slightly bifid or notched at apex.
Palpi short thick joints — paraglossa
oval rounded.

The fore coxæ are globular
in round sockets — therefore this
genus is not belonging to the
Prionidæ according to Spinola's
definition   Feby 12 1854

Strategus Alceus ?. ♀.

The labrum consists simply
a a pair of divergent pencils
hair attached to the undersurface
clypeus.   The labrum also
inner side forms a pair of
divergent pencils of hair   the tips
which project beyond the edge
the corneous part upper edge of
mentum.   The maxilla has 7 or
regularly sized teeth.

Chefonarium

A genus of curious Coleop.
evidently allied to the Byrrhi
but I cannot find described.
legs all contractile & lodged
in oblong cavities, so as to be
plane with the undersurface
of the body; tarsi 5 jointed
of which the 3rd is prolonged
beneath into a flap & 4th is
minute, simple.   the fronts
of the protho. projects as a
rim beyond the head.   Head
lodged in a cavity on fore-
breast, split when retracted.
Eyes large prominent, very
smooth.   Antennæ very
curious   the basal 3 joints
very large, flattened & applied
close within a central channel of the
fore & mesosternum   so that
the mouth is entirely concealed
beneath, — the remaining jts
are very slender, elongate.
Most often the antennæ are
found broken & I am not
sure whether the sketch as
above is perfectly correct
March 10 1854 — the body is
oblong, but slightly convex
— colour black or castaneous
nearly smooth.   But one
species is variegated with
golden pits in oblong streaks
& patches similarly to some
Byrrhi.

---

very closely resembling in
its buccal formation   the
Trachyderes succinctus.
Dissected & figured No 12
Feby 27 1855

Erotylus —   the mentum is elongato-
quadrate with the front in its middle
forming a triangular plate projecting
forward from base to apex — the upper
corners of mentum are acute: the whole
mentum horny & solid: there is no
intermediate membranous piece adjoining
upper edge of mentum, the whole of the
labial parts being very solid & horny — the
basal support of palps is horny & furrowed
rather projecting in front & concealing the
root joint of palps — the ligula-paraglossa
is horny & closely adhering to supporting
piece of palpi: —
16 Octob. 1855

There is here as in Lophonocerus
&c. a horny piece between lip & palps — in front, below
bottom of cleft of lig parag., from which arise a few
long hairs — a x ...

Geodephaga Brachinidæ

No. 25

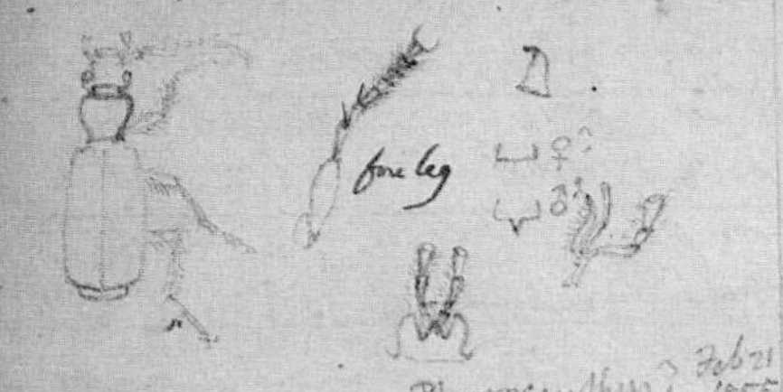

Pleurocanthus? Feb 21 1855

Group of Galeritina (judging from its structure tho' I think it is called Helluomorpha, group of Graphipterina dusky blk, short silky brown pubescence — limbs & mouth longer pubescence — habits same as Galerita; viz under rotting logs & roots)

Hd. rather enlarged behind eyes, but not prolonged, contracting abruptly into a broad neck. Spines of fore tibiæ long & strong. Tarsi of fore legs same as those of other legs, viz ~~it~~ 1st jt. rather the longest & with 2nd & 3d triangular ~~fourth~~ 4th semicircularly sinuated. densely pubescent, pubescence coming along on sides, & underside. claws strongly curved, simple. Palpi with termnl. jt. ~~ab~~ increasing in breadth towards apex which is truncate the apex forms a circular cavity with a membrane within; hers, the difference between this structure & that of Galerita is that in the latter the tip is flattened off the apex is the sharp edge of one side of this cavity the other edge being lowered & the true membranous apex is on the ~~lower~~ inner surface of a spoonshaped joint. Thus in using the palpi as tactors the Galerita applies them *flat* on the substance — the present insect only touches it with their apices, like the majority of Insects. Being a rare insect I have not dismembered the specn. & therefore not investigated the ligula April 10/1853

Geodephaga Brachinidæ

No. 26

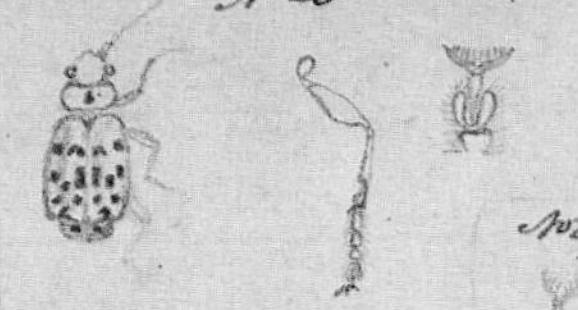

No. 27

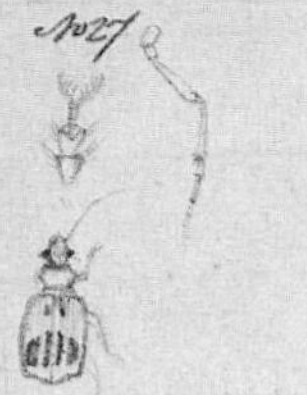

No. 28

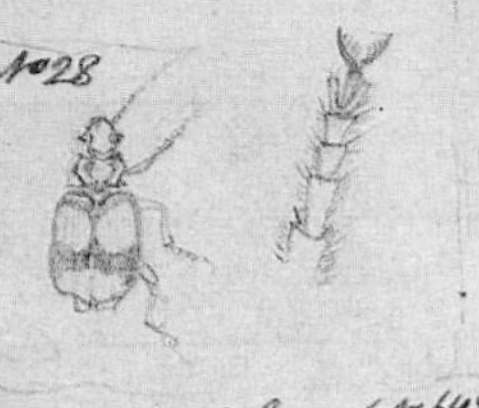

No. 26 — Lebiina (No 649) — Chelonodema Lap. — tarsi with triangular jts. penultimate deeply bilobed. claws thickened at base as in Agra &c. & pectinated. Felt underneath fine cilia at margins.

This sp. peculiarly arboreal in its structure. I found it, clinging to a leaf.

No. 27 Lebiina (No 487) 9.
tarsal jts. rather elongate triangular penultimate scarcely perceptibly sinuated. claws not thickened at base, pectinations very thin & wide apart — tarsi coarsely spongy pubescent no felt
This sp. less arboreal, I swept it from low plants on campo

No. 28 Lebiina (no 609)
tarsal jts. elongate subtriangular penult. deeply bilobed, lobes not enlarged & rounded at apex, claws very little thickened pectinations rather sparse
running on leaves.

Geodephaga Brachinidæ

No. 29

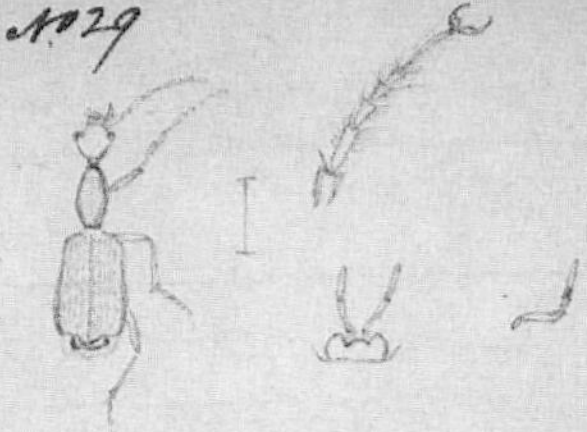

Casnonia (blkish brassy, polished tho. transversely wrinkled, elytra with rows of regular shallow punctures) beaten off of bushes. Head regularly rounded behind & gibbous, neck very narrow &c. — Evidently more allied to Agra Callida &c. than to Galeritina or than ~~the~~ last are to either Casnonia or Agra. Palpi all with term. jt. elongate, cylindroid, narr. at apex the extreme tip truncated. the parts polished as in Agra &c. & not pubescent as in Galeritina. claws simple with a dilatation within near their base (April 12/53)

Fine Dryptide new Ega, in leaves-folds habits of Agra

Note — on Tarsi of some genera

In Casnonia the 1st joint of the tarsi is shorter than the 5th, & also than the 2nd, 3rd & 4th together. the 5th or claw jt. being the longest jt. in the 3 pairs of tarsi.   In Galerita the 1st jt. is equal in length to the 2nd 3rd & 4th together & double the length of the 5th or claw jt.   As the two genera both belong to the terrestial — cursorial groups of Brachinida this difference is important. In the arboreal both leaf-clinging (Agra) & leaf-running (Lebia) — groups the 1st & 5th jts are about of equal length & less than the 2nd 3rd & 4th together — But in Ctenodactyla, the 1st is much the longest joint, a strong proof that it is really far removed from Agrina & perhaps an arboreal, transformed form, allied to Anchomenina.

Ctenodactyla Chevrolati 4d. narrowed in a straight line behind the eyes, not forming a rounded gibbous cheek, as in Casnonia, Agra &c. Spines of fore tibiae very small, as in all arboreal-tarsi Geodephaga. Tarsi although dilated &c as in the arboreal families Agrina & Lebiina, yet offer some points of difference.   the 1st jt is longer than 5th & about as long as 2 3 & 4th together. the 3rd & 4 jts have much more acute angles, in Agra &c. they are more quadrate with round angles.   The claws are pectinate but in a very different manner to the Agrina. they are straighter, are not thickened, & the pectinations are shorter, more like the teeth of a saw, the points of the claws & their claws being more acute. — the penultimate is bilobed & covered beneath with glossy felt as in the Agrina.   The palpi have their terminal jt. oblong, rather suddenly narrowed at tip, the tip itself being truncate or blunt. the 3rd jt. is about as long as the terminal — which is a character of the Anchomenina.   This species is found on the Indian corn, concealed in the bottoms of the shrinking leaves in co with Leptotrachelus 1020, to which it has the closest resemblance, excepting solely the sensation of the claws. I cannot conceive how Laporte could make it a subfamily with Agra, it is rather a form allied to Anchomenina. April 12/53

Anchonoderus Reiche?

I select this as an illustration of tarsi of the cursorial type — the 1st jt. is about as long as 5th & as the 2nd & 3rd together — chief difference is in the length & strength of the tibial spines especially of hindmost legs. April 12/53

Loxandrus?

Ligula elongate 4 dentate Paraglossa thin membranous not projecting considerably beyond front edge of ligula fore tarsi ♂ 3 jts dilated obliquely cordate — Mentum with a simple triangular tooth in the notch.   Novem 16. 1854

OLR. lozenge shaped and spot across sutures on tip. shape of Calathi but tarsi unique & simple under stones & logs

hind most tarsus

fore tarsus much magnif:

labrum

mand.   max.

mentum, lab. palpi & ligula (behind)

abdomen upp. side.   elytron

labrum

mand.   max.

elytron abd.

fore leg ♂

hind leg ♀

mid leg

tibial spine hind leg

labium

fore tarsus

tibial spine hind leg

hind most tarsus

Calleida ? (cont?. green-bronzed sp. Saut.)
— Hd. not restricted behind into a very
narrow neck. Spines of fore tibia not
apparent, & are very minute on the
other legs. Tarsi same to 3 pairs of legs
as drawing, densely pubescent beneath,
penultimate only clothed with close felt beneath
Claws much thickened at base so that the
points of the pectinations come nearly in
a straight line. The term. seg.
of abdomen above is very different
from the covered surf: being horny
blk. polished & with a few punctures.
— Query if the truncation of elytra
is not connected with the movements
of apical seg. of abdomen in
sexual intercourse? It is rather
singular the ♂ in Brachinidæ
generally are not distinguished
by dilated tarsi (in those genera
where all the tarsi are simple)
as in nearly all the families of
Carabici. In Mandibles, paraglossa
&c. there is no resemblance to the
Cicindelidæ, no more than there is
in the other fam. of Carabici.
April 8/53

not pubescent.

Galerita   **unicolor** Latr.?
Hd. constricted in a narrow annular
convex neck. Spines of fore
tibia long & strong, especially the
notch spine; hind leg long tibial
spines. Tarsi of fore legs, have
the 3 first jts. dilated into a lobe
on one side, beneath ridged &
furrowed & densely pubescent.
of hind legs not dilated 1st to
4th jts. gradually decreasing in
length, 4th jt. with angles rather
prolonged & furnished with
longer hairs. Beneath they are
also pubescent & spinose.
Claws strongly & regularly
curved quite simple.
Mentum has no tooth in
its notch (but this a character
of smally importance) all the
palpi hatchet shaped terminal
jt. the edge ground fine &
rather membranous. pubescent.
the ligula appears the same
shape as in Calleida
paraglossa I cannot detect.
April 9/53 — The above ♀
find is a ♂ — since found a ♀
the fore tarsi are exactly as in
N°25 — being mm slender, 1st jt.
elongate, 2 & 3 triangular, 4th rather
deeply & angularly sinuate. they
after 11/53 do not present the
ridges beneath

Agra (white legs blk. knees.)
Hd. prolonged, inflated behind eyes, &
form?. a very narrow annular neck.
the spines on fore tibiæ not
apparent — at apex very minute
— also very minute on other legs.
Tarsi differ only in the
fore legs being rather shorter & broader
than in other legs. In all the
penultimate jts. densely
felt covered beneath, the others coarsely hairy
usually in edges, giving a palmate appearance
to all the feet. Claws very similar
to Calleida. The blade of max.
small & thin, the claw distinct &
almost appearing moveable like the
Cicindela. max. palp. thick, oval jts.
inner — palpi — term. jt. of length of 1st jt.
& squarely truncate. term. jt. of outer
palpi also squarely truncate.
Labial palpi large in proportion to max.
jts. solid, horny, glabrous, the palp
jt. not having a fine membranous tip
as in Galerita. Mentum not
plane, the side angles entering inward
towards mouth, base of lab. palpi
little appearing beyond edge of the notch
of mentum. The tongue as in Cal
but my dissections are seldom sure
with this organ. Mandibles long &
whole insect glabrous. April 9/53

Papilionidæ

N°1

N°1
a
apex
b

N°2
c

fore leg
♂

P. Sesostris

1 costal nervule
2 sub-costal nervure
3 1st subcostal nervule   ? better called branch
4 2nd Do Do
5 3d Do Do
6 4th Do Do
7 upper Disco-cellular nervule
8 middle   „
9 lower   „
10 Median nervure
11 - 12 - 13 median 1st 2nd & 3rd nervules
14 sub-median nervure or radial nervure
15 (hind wing) anterior margin

16 precostal nervule
17 subcostal nervure
18 subcostal 1st nervule
19 2nd   „
20 upper Disco-cellular nervule
21 middle   „
22 lower   „
23 median nervure
24 - 25 - 26 1st 2nd 3rd med. nervules
27 28 upper then may be
called the median
outer nervures

N° 1 Papilio Agavus. a transverse nervure connecting median with hind nervure of
fore wings — showing a wide difference between genus Papilio & the Pierides. Disc. cells all
closed by perfect nervures. fore legs perfect, ungues long, simple, approximated. fore tibia
a fleshy process on inner side. — Feby/52
N° 2 c Papilio [strikethrough] ♂ — hind leg of ♂ showing enlarged tibia — the enlargement is
angular.

Argynnites

No 1

No 2

No 3

No 4

No 5

no 6

no 7

No 8

No 9 — see next page

No 10

No 2 fore leg

No 8 — Agraulis

common object in old order — under side buff, with four semicircular spots of what is suffused

precisely similar in antennæ & palpi to Argynnis No 1 — elongated wings — disc.ᵃˡ cell closed by an almost obliterated suture (not tubular nervure) —

fore leg is hairy like No 1 ungues of other legs are short & have black palms between them

No 1 Argynnis   the discoidal cell of fore wings closed by a nervure touching the median nervure much above its second branch — palpi hairy on all sides, except extreme tip

No 2 Argynnis   the disc.ᵃˡ cell of fore wings closed by a nervure touching the median nervure at the points, or very close to, its second branch — palpi less divergent & covered with longer hairs than No 1, terminal joint minute nearly concealed in long pile of 2ⁿᵈ joint.

No 3 Melitæa   disc. cell of fore wings not closed, the faint suture from branches of the costal nervure not touching the median

No 4 — side of No 1 — ungues long — palpi bent forward, proboscis long, fore legs very hairy like No 5 fore leg of Melitæa not hairy like that of No 4 — 6 midlg of No 3. ungues short not palpi of not bent

No 1   Heliconia   No 2   No 3   ♂ No 8 a

No 9 Eueides Domklia   Lycorea   No 8 b

No 8   9 a

Mechanitis   Ithomia

No 4   no 6   fore leg ♂ Mechanitis   No 10   10 a ♀

fore leg ♂ of Saio ♂○—

No 5   No 7 fore leg ♂   hind leg   Ithomia Sao

—According with No 1 in palpi; neuration eyes; foreleg of ♂ &c. ab.t 15 species according with No 4 2 sp.

No 8 &♀ ♂ larger — the pencils of hairs of tail retractile — when exerted strong smell — ant. more clavate & shorter than ♀ of th

No 1 Heliconia Melpomene   discoidal cells of both pair of wings closed — palpi very similar to no 8 antennæ & neuration of wings widely distinct clothed thin long pubescence

No 2 side view of head of Do. ~~antenna~~ palpi much bent forward, divergent apices — tarsal joints conglomerate at apex with 2 short spines, 5 ca-joint

No 3 fore leg of Do ♂ slightly hairy not ~~so~~ densely pilose like Argynnites Nos 1 & 2, ♀♀ terminal

XX — The above No 1 agrees with 7 other species of Heliconiites wh. I have examined some species differing in more slender club to antenna & abdomen much longer than anal edge of hind wings.

XXX Heliconia very bare of pubescence of head, palpi; head abd. under of wings

No 9 & 10 in page Argynnites — show tarsi & ungues of (No 9) Argynnis No 1 & (No 10) of Heliconia Melpomene — No 9 the ungues are long, simple, free of the spongy palm No 10 they are double, small curved & have the spongy palm between. No 10 is the character of Argraulis, Heliconia

No 9 small antennæ ♂ ♂ ♂ ant clavate like Argraulis. to wh. allied also in origin of 1st front branchlet. Disc. cells closed all by tubular nervures. No 10 Ithomia neuration nearly identical with No 4 fore leg ♀ spinose tarsal apices like Nymphalidæ

XX larg. diaphanous tt. long claw & bd. wings & ante. like No 4 but 2nd pair much thinner as the junction

No 4, 5 Heliconia, Mechanitis hd. small, palpi term. joint minute; fine th. 1/3 length abd. abd very long clavate. hind fore margin narrow, pencil of hairs to ♀ some if under fore pair — No 7 fore legs, tarsi abortive — ♂ hd. small but sub

the ♂ of No 9 has not the pencil of hairs on fore legs & hind wings

Belemnia Eryx of Authors

Euchromia

closely clothed wing
Zygænide? very beautiful metallic-colo
sp?, Hd. tho. & base abd. beneath, 6 spots
on tho. above, pale silvery green. tho. blk
—abd.—4 basal sgts. on sides pale silvery
blue — 3 succeed'g sgts. same parts replaced
by dark velvety blue — beneath velvety
crimson. fore wing, basal third golden
—rest black with a short belt red, across
at 2/3rd length. Hind wing blk.——

Neuration — nervures strong - costa horny
— the discoidal nervures may be considered
as 3 — thus the lower is joined to median
very close to 5th median nervule — the middle
originating at end of cell & getting with the
discocellular nervule — & the upper arising
from subcostal nervure at a distance
beyond end of cell. This will leave 2
nervules besides terminal fork to the
subcostal vein. 1st emitted at a
distance before end of cell, 2nd shortly
after end of cell — & the terminal fork
taking place nr. apex & costa of wing.
This is a better way I think of describing
this form of neuration, than considering
the upper discoidal nervure as a
branch emitted from lower side of
subcostal nervure & not as a
discoidal nervure. — In the neuration
of Lepidoptera it appears that a
nervure may arise where it is wanted
owing to another nervure being
employed in another direction, but this
replacement of organs is of course
according to a system & the system
we have to find out. Thus, in the
present case the lower discoidal
nervure is so low down as to be
brought closely into connection with

the median line of nervures, & in consequence
another nervure was wanted above, which
is = the upper discoidal, here so considered.
——
        The disco cellular is long, oblique
& imperfect for a slight portion towards its
junction with the lower discoidal.
—   The palpi are formed of 3 short thick
jts. very distinct, nearly of equal
size, curving to the face, the tips of
the 3rd jt. just visible from above.
The maxillæ are strong & as long as
the tho. The antennæ are thickened
towards 2/3rds the length. — bipectinated
the pectinations longer at the thickened
portion & ceasing towards 3/5 ths the length.
The legs & claws are strong, the postvillus
small. The Tarsi with a double row of
short spines beneath. the tibial spurs
long & strong on mid. & hind legs. The legs
are free of pubescence, having a coating
of very closely applied scales
Nov. 13 1853

N°6

palpi

antenna thick
slightly pubescent beneath

hind leg
tarsi tapering
claws very
small

Trochilium (size & colours of T. tipulæform)
—Neuration from wing extreme narrowness &c
very peculiar. The cell terminates
ab¹. middle of the wing at a point
where the texture & appearance of the
wing appears thickened & obscure
through which appears very indistinctly
an imperfect entire transverse across
end of cell. The subcostal nervure
after throwing off its first nervule
(considerably before end of cell) becomes
indistinct. At the end of cell it
throws off its 2nd nervule & then
becomes quite obliterated. The 3rd
emerges from the thickened portion at
end of cell; & what appears to be truly

the continuation of the subcos
nervure, arises from the same
clouded part, below the origin
of the 3rd nervule, this form
a bifurcation about half way
between end of cell & apex of
wing. The lower fork of the
bifurcation I consider to be
the same nervule as that
marked — a — in the fig. N°5
wh. I have called the upper
discoidal nervure. Thus
considered the other remaining
nervures become intelligible
— the middle & lower discoidal
nervure arise from the
middle part of the extreme
of the cell, one below the
other & run parallel to
ea. other & also to the subcostal
nervure & the terminal median
nervure. This latter arises
a little below the lower disc
& is disconnected with the
median nervure, which then
appears with only two nerv.
— Thus we see that the
system of neuration here
has a point
of concentration, as it were
in the obscure transverse fin
(or fold) in the membrane, perhaps
crossing the end of the cell. He
the secondary nervures of the
wing all originate without
any continuous connection
on their part, as tubular vein
with the base of the wing. T
is no post-median nervure
fore wing. — The Hind wing
being more ample has a sim
neuration to other allied Lepid.
subcostal nervures however ru
close along the costal edge. The
median is as usual & a disco
runs from median to disc. nerv
wh. is rather indistinct as it
neur. junction with subcostal.
Nov. 14. 1853

No1   1a   No2   No2a

?Euchromia
1b
c
d

No3

Glaucopis

Macroglossa with half abd.
Neuration very similar to Sphinx
subcost nervure emits 2 nervules before
end of cell. an equal bifurcation at
tip. abt ½ way, between end of cell & apex
thus 1 nervule less than Hesperia —
upper discocellular none — middle slanting inwards strong & tubular
lower tubular but more feeble
than middle — all nervures strong
& costal edge horny — palpi very
densely clothed with fine downy scales
close broad at apex concealing apical
th. & applied so closely to face as to
appear a conical beak
proboscis reaches to abt 3/4 length of
body — Jan 7/53

No3
b   c   d

3a

?of Zygana?   The chief diff. in neuration from Diurnes is the extreme right branchlet of frnt
nervure, which becomes isolated in Lycænidæ & Hesperiadæ, here joins the median nervure — the 2 nervures being
connected by a feint entire which forms the disc. cell. Feby/52

No2 Trochilium — neuration same as No1 — palpi longer- term. joint more pointed. there is an
incipient recurrent nervure rudimentary at the point of emission of first right branchlet of frontal nervure
which is connected with the feeble line connecting with a similar recurrent nervure of the median Feb1/52

No3 Sphinx — the nervures are very strong on front edge. Frontal nervure wants
the terminal branchlet — the extreme right branchlet is connected with frontal nervure as
in Diurnes & not as the Sesiæ & Erebites. Disc. cell closed by tubular nervures — in fore wings
the connecting nervure between median nervure & frontal is translucent not brown, hornys
April/52

No 1

1 a

b

c

d

Noctua
(Orthosia?)

N.º 3 ophideres
April 27/53

Allied to Graphiphora segets? nigrum
whilst
homoptera
with corms
♀
palpus

No 1 — Erebus — the neuration less resemblances to Diurnes, than Attacus has. the median nervure has an additional branch, terminal. the additional cell connected with frontal nervure remarkable. basi elongate, spines long &c. March /52.

No 2 Noctua — Orthosia? (a common drab coloured uniform, few markings) Neuration almost identical with Erebus — ♂♀ antenna quite simple — legs without fascicles of hairs. palpi middle jt. densely scaly in front, more dense towards apex. term. jt. nearly naked, oblong or elongate pointed, the points approximating. In neuration the anastomosing of the subcostal nervure with its nervules, forming then the 4 d rate subcostal cell is remarkable. In the drawing the costal nervures all are brought too much into the middle of the wing, to make them more distinct, in nature they are all very close to costal edge the tibiæ tharsi have a few short prostrate spines. March /53

No 3 ophideres (comm.) neuration differs little for Erebus. the subcostal is nearly straight to the emission of upper disco-cellular — whereas in Erebus it bends down into an angle to the upp. disco-cellular. the perfect insect has its upper wings closed over lower roof-shaped, but Erebus is expanded — larva fan

Bombycidæ

No 1

No 1 a

b

c

d

No
♀ ?

Hepialidæ ?

midleg

hindleg

No 1 *Attacus* — Comparing with Diurnes (Morpho) the front nervure wants 3 frontal branchlets & divides into 2 main branches near its base — the disc. cells all perfectly open. — no perceptible proboscis. legs short short woolly. Tarsi thick — frießart (6) 4th tarsal joint has 2 short spines — hind tibiæ appear to want the median pair of spines — the ungues are short & hidden in the fur of feet — have palpus. March/52

2. *Hepialidæ* ? / whitey eating. brown tail (♀? depositing much cottony matter by tail — flies hovering with pausy outputs in evening. The neuration widely different from Attacus — the great difficulty is in considering the discoidal nervures, there appear to be three! the 3rd joined to 3rd median nervule — the first originates at a minute upper discocellular, in the usual place — the 2nd is isolated — the middle & lower disco-cellulars are indicated by # nearly obsolete lines, curving angularly inward, the middle from a short rudimentary recurrent nervule — the lower from a similar rudiment below. the palpi are short, very thin, nearly naked, pointed the legs are long, slender. May 4/53) The neuration wd be same as Noctua if the middle (#) discoidal nervure were continued by a recurrent nervule to the subcostal nervure, thus forming the Quadrangular subcostal cell. May 5/1853 —

Geometridæ

Nº1

Nº1 Bradypetes allied? neuration differs
from Divines, in the furcations of the subcostal
nervure - it here, forks in 2 equal divisions abt
the middle distance between end of cell & apex of
wing - the posterior fork is simple, the anterior
throw off 2 nervules to the costal edge - The
first subcostal nervule is emitted much
before end of cell touches the costal nervure
& blends also with subcostal nervure after
all the furcation. As to the rest the
neuration is similar to Divines - the upper
discocellular is short & slants interiorly - the
middle & lower do are transverse across wing of
equal length, & rudimentary leaving lower discoidal
nervure as in Hesperia &c isolated - the
lower wing want the lower fork of the
subcostal nervule (homologue of lower discoidal
nervure in fore wings) -
The nervures are all feeble - like the
greatpart of Erycinidæ -
palpi very sparsely scaley - curved divergent
in middle, approximating towards apex - apical
jt minute naked, visible from above - proboscis
regular - legs as above - spurs beneath at
the knee. ant. scarcely pubescent
Feby/53

Urania Leilus June 8 1854

The tailed, white moth much resembling Ourapteryx sambu-
caria - Pará, Ega &c. flight slow settling on leaves wings flat
extended. - I had thought it a Geometrid, but its characters
are those of Bombycidæ - Its wings are clothed above
with short cottony & downy scales - Compare the neuration
with Attacus - the costal runs along to near the
tip of wing - there is no cell properly speaking, there
being no rudiment of lower discocellular. Note the cell
bearded upwards of median near its origin - same as
in Attacus - The proboscis is null there are 2 folds
separated - they are shorter than the labial palpi, inside
finely fringed, upper edge with 5 or 6 erect papillæ? or
is it dotted pencils of hair - scales? - The Antennæ have about 40
pairs of branches one apparently to a joint, the branches pubescent
November 8 1855

Bombycidæ

Saccophora Batesii (Newman)

Larva — enclosed in a fusiform case
of leaves, so well woven together that it
is difficult to detect the sutures. abdomen
swollen, flattened behind. Head & 1st seg. thorax. Head rufescent 2nd & 5th segt tho. yellow
with 2 black stripes. rest of body pallid
yellowish or olivaceous, sprinkled or marbled
with dusky atoms.   whole a stripe of
yellow along whole body below the stigmata.
Beneath hd. Thoracic segt. & 1 & 2 abdominal
segt. black. rest of abd. ruddy flesh
colour. 10 pairs of false legs. viz. on 3 & 6
abd. segt & anal segt.   a few long
white hairs on hd. & fore part of
body.   the case is found
attached by one or two strong
silken threads to leaves of Melastoma
Byrsonima & also to grasses. —

Passes transformation within its
case, ♂ & ♀ both winged. ♂ differing
in branches of ant. rather longer
— a pair of pubescent processes to ea.
jt. of ant. I count 39 jts in ♂
& 43 in ♀ — to the 23rd jt. they are long
thence suddenly shorter to apex, this part
of ant. being curled.

103 — Neuration very unlike Saturnia (Attacus) No 1. The costal nerve is faint & terminates on costa
the 2/3rd length of wing. Subcostal throws off it. 1st & 2nd nervules before end of cell, the 2nd nervule conveying the
strength of the subcostal neuration. The upper discocellular nervule is very short. the middle is of same length
as lower, the middle null in its centre, the lower perfect. The neuration resembles that of Sphingidæ
the hind legs has no mid-tibial spurs (? !) —   No perceptible maxillæ

Bombycidæ —   satiny–pearly–white –with
transverse belts of denser scales across fore wings—
–antennæ bipectinated   pectinations not very
long & nearly equal in siz. to the apex, &
all ciliated.   palpi   very short & feeble
proboscis entirely wanting.   on comparison with
the neuration of *Saccophora*. with difference is apparent. —   Ega 27 Jany 1856

Bombycidæ — the antennæ have very long pectinations almost
immediately from the base & they finish abruptly a little before the
apex, leaving the end setiform. the proboscis is moderately
long. the palpi have the basal jt. much the longest &
curving up the front & furnished with very long hair–scales
the 2nd & 3rd jts. being short & nearly naked
                                        Ega 14 July 1856

Chelonidæ". (wings sooty-blk, fore pair
a short transverse belt, hind pair a
a broader central band across, ochreous
yellow, - very convex inocat, settles upon or
under leaves wings - fore pair covering hind
pair - day flyer - _____ body beneath
white - abd. above a dorsal row of
ochreous yellow spots.)

Neuration - costa horny - costal
nervure terminating on costa at 2/3 ds
length of wing. Subcostal emitting its
1st & 2nd nervules before end of cell, the
1st much before the 2nd, & inclining
downwards at a very obtuse angle
after the emission of each. 3rd & 4th
nervules forming the final bifurcation
which is nr. apex of wing. The 1st
discoidal nervule arising from subcostal
abt. 1/2 way between end of cell & the
bifurcation of subcostal. 2nd do
arising from a short upper disco cellular
- 3rd do arising 1/2 way between 2nd
& median nervure, having no immediate connection with the latter -
middle & lower disco-cellulars
of abt. equal size & imperfectly
tubular. 3rd Median nervule forming
a considerable angle to receive lower
discocellular, at a short distance
beyond after its bifurcation from 2nd
nervule. +
Hind wings - costa straight as also
costal nervure, terminating nr apex
call angle, the 2 disco-cellular
nervules straight, transverse, faint
but tubular, _____ both of equal
length, the upper arising from subcostal
at a long distance before the emission

of the nervule of the latter & the
lower joining the median at
a short distance before the bifurcn
of 2nd & 3rd nervules of the latter.
head small, very short in
elevation, palpi very strongly
curved, the points reaching backwrds
to the crown: basal jt. like the
others covered with short scales.
terminal jt. a mere obtuse
conical point.

Antennæ, elongate, very slender
setiform - composed of minute
joints, beneath with a very
short - scent pubescence.

This insect closely allied to
the genera Callimorpha, Euprepia
Euchelia &c. is classed by the
Germans with the moths
(Zygaenae?) figured here No 1 - 2
5 & 7 - as the same family
Chelonidæ. - I find the
neuration different in a very
essential point viz. the separation
of the lower (or 3rd) discoidal
nervure from the median
nervure. They agree in the
presence of a 1st discoidal
- but differ again in the position
of the first 2 subcostal nervules
in the present insect they are
both emitted before end of cell
- in the others only one arises
before the end of the cell.
Nov 20 1853
_____
Sep 9 1854 At Villa
Nova, found the pupa
of the present species
- It changed about 2
hours after I received
it - _____ suspended by
tail to underside of
leaf - the chief peculiarities are
the legs protrude in form of
short pointed tubercles - & it
has a bifurcated branched
horny process at apex of
head

Chelonidæ? (wings deep black,
a yellow-orange band completely across
fore wings, a little beyond middle) flight
random, in forest, settles on leaves, wings
overlapping to form △ figure.

Neuration - costa horny - subcos-
nervure emitting 1st nervule 1/2 way
between base of wing & end of cell: it
nervule 1/2 way between end of cell &
apex of wing - 5th & 4th nervules form
final bifurcation very near apex of wing
- the 4th terminating on margin below
apex. the 1st discoidal nervule befor
from subcostal nearer apex of wing
than end of cell. 2nd do arises from
a short upper disco-cellular. 5th do
midway between 2nd & median nervure
having no immediate connection with
the latter. Middle discocellular nr
obliterated; lower, faint. 3rd Media
nervule forming a very slight angle
& near its origin, to receive lower
discocellular. _____ Hind wings, for
edge slightly flexuous, neuration
precisely as in No 9.

Palpi, basal jt. large, same length as 2 -
terminal shorter, obtuse. The whole of
it nearly, appearing beyond the face
viewed from above, & drooping. An
thick nr. base, then gradually taper
to apex, formed of very small jts. covr
scarce perceptible pubescence
Nov 24 1853

This sp. & its allies are nocturnal
resting found in repose on leaves

patch
of brown scales
& slates
above margin
little fan

wings extended
in repose

♀

Euchromia or Glaucopis? — Ega

Euchromia? = neuration
of fore wing is on essentially
the same type as that of the
Glaucopes: i.e. the middle
& lower discocellulars forming
a re-entering angle whose
apex is traversed by a
longitudinal
central fold, & the lower
discoidal joining to it in
origin with the 3rd median
Here the post median fold
exists. . . . . . It is the
hind wing which offers
notable difference. The
abdominal margin is ample
has 3 folds closing like a
fan; the costal edge
produced into a broad
angular expanse — the
discocellulars are straight
transversal or nearly so.
here is also a singular
al patch of dense scales
middle of discal cell
The palpi & proboscis are
early identical with Glaucopis
antennae are similar
so. They are elongate &
being — & being a ♀ (Joutfin)
joint are bi-serrate, i.e
it has a sharpe point on each side
which rises a curved seta
25 August 1856

neuration in dells in forest, settling very low
beneath leaves. they fly. like all Glaucop
Neuration rather different from
any hitherto dissected.
Fore wings has a tubular, regular
discal nervure arising at base
cross passing through the discoidal
cell & joining the median nervure
a little after the origin
of 3rd median nervule, at a
little before its junction it makes
a short curve & to the
this curve joins the discocellular
nervule which is obsolete except
the portion near its junction
with the discoidal nervure &
upper discocellular.
Hind wings have no discal fold
nor discal nervure the discoidal
crosses the end of cell obliquely
joining the median nervure
between the 2nd & 3rd nervules
the median forming a slight
angle to receive it. —
The palpi & legs offer nothing peculiar
the anti. in ♀ are slender &
furnished with very short projecti-
ations regular in size from base
to tip
Ega 5 October 1856

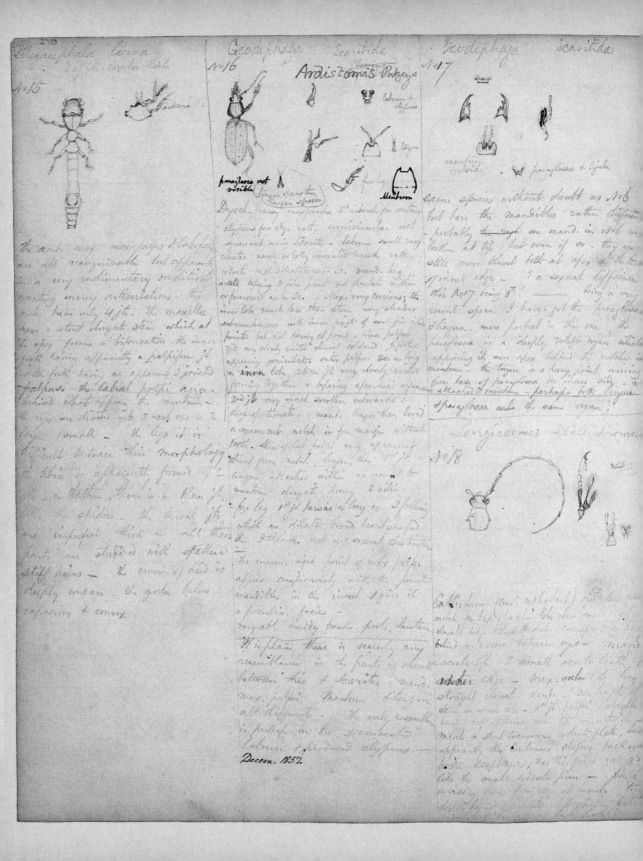

20

Megacephala larva —
of M. carata Bonle?
No 15

Geodephaga   Scaritida
No 16
Ardistomus Pilcarp
labrum & clypeus
tongue
paraglossa not visible   tongue of another longer species
forleg
Mentum

Geodephaga   Scaritida
No 17
mentum outside   paraglossa & ligula

The ant. max. maxr. palpi & labr. palpi are all recognisable but apparently in a very rudimentary condition wanting many articulations. the ant. have only 4 jt. The maxillae have a stout elongate stem which at it apex forms a bifurcation. the inner fork having apparently a palpiform jt. outer fork having an apparent 3 jointed palpus. the labial palpi arise behind what appear the mentum. & eye are divided into 3 each one = 2 larger, 1 small. the legs it is difficult to trace their morphology. the tibia is apparently formed of 2 — or rather, there is a knee jt. like the spiders. the tarsal jts. are imperfect, thick — all these parts are studded with scattered stiff hairs — the crown of head is deeply concave, the gula below capacious & convex.

Dysch. Maxy northorder 3d interch for anterior clypeus fore edge rather. semicircular not emarginate as in Scarites. labrum small very ciliate nearly wholly concealed beneath rather 4 dents not 5 dentate as in se. mand. long acute tapering to fine point not dentate within or furrowed as in se. — Max. very curious; the inner lobe much less than stem. very slender submembranous with inner fringe of small fine cilia pointed but not horny at point. inner palpus jts. very closely united almost soldered. together appearing geniculate. outer palpus 3 as long as inner lobe, stem jt. very closely united forming together a tapering spearhead appearance 2nd jt. very much swollen outwards & disproportionate. ment. longer than broad a square cut notch in fore margin without tooth. stem of lab. palpi very apparent. arise from notch, longer than 1st jt. tongue attached within as usual to mentum elongate horny 2 cilia forleg 1st jt. tarsus as long as 3 following which are dilated broad heart shaped the 9 I think. not a sexual distinction —
the curving rigid point of max. palpi appear conspicuously, with the pointed mandibles in the insect & give it a peculiar facies —
very abt. muddy banks pools, Santiago.
It is plain there is scarcely any resemblance in the parts of mouth between this & Scarites. mand. max. palpi mentum & tongue all different. the only resemblance is perhaps in the semiconcealed labrum & produced clypeus —
Decem. 1852.

same species without doubt as No 6 but here the mandibles rather different probably the right one mand. in No 6 was broken at tip — but even if so — they are still more blunt both at apex & the inner edge — ? a sexual difference this No 17 being ♂ ? —— being a very recent species I have got the paraglossa & ligula more perfect in this one. the paraglossa is a deeply cleft organ ciliated appearing its mere apex behind the notches in mentum — the tongue is a horny point arising from base of paraglossa or more likely attached to mentum — perhaps both tongue & paraglossa are the same organ?

Longicornes   Callichroma
No 18
Callichroma (sens ?) ... notch ... behind a crown between eyes — mand. acute tip 2 small acute teeth inner edge — Max. ... long straight ... slender ... stem or inner lobe — 1st jt. palpi slightly ... mentum ... mentum a stout transverse ... plate ...
Decem. 1852.

No1

dorsal
3 cutters
1 molar/tch

See P. 45

No1 f

No1
a —
— b
c —
d —
— c

Acridiida

epistoma
labrum

tongue
appendx

hind tarsi, all the
joint vesicular
the pulvilles a larger
plantar vesicle

Jan 4. 1853

Acridium (comm grassgreen species, bluish wings, slender, found on herbage on the beach chirterous) I have dissected the lower lip with particular attention. a is the cavity from which the prothorax was detached from the head. b is a narrow piece which I suppose is part of the integument of the head, but as it is coriaceous like the parts of the mouth it is not well distinguishable from them; it happens otherwise with Hymenop. & Coleop, where the general integument is so firm & hard that the whole margin of the buccal cavity is well easily defined. c is the base of lip-palpi immovable. d is a hydrate leathery plate near base of anterior surface of which arise the lip palps - query mentum? e is a pair of soft leathery plate fixed by their bases to d, naten to its inner surface, query homologous with what are called lip palps in Libellulides? (See organ h No Libellulida) if so d would be homologous with organ g called mentum in same family (organ i bis) f shown the base or insertion of the maxilla. the max. apex has 2 sharp teeth spines - the hood is leathery & protects the stem, regular oval concave within. the mandibles are short very strong & thick pieces, brittle too long, concave convex, upper edge spirit tubercles. the tongue is elongate 2 keiled, attached by membranes to the lining of esophagus & to form lip?

Feby 1853   The large comm locust — shortly distinctly the insertion of the lip palpi — It appears we may consider the piece o as homologous with mentum. max dept relieft at tip, no spires 3 tubercled —

No1 **Blatta gigantea** Lin. wings extended elytra horny-membranous veined. hind wing membranous — posterior p (at point g) folded fan form, anterior p forming one fold over post p in repose. abdomen flat depressed the junction of upper & lower arches of the segments meeting in lateral edge where the upper arches slightly project over lower. labrum a - advanced covering mandibles, but retractile in mastication. mandibles b short thick, many toothed. maxilla c with inner lobe fleshy stem 2 toothed at apex & fringed within side fleshy d tongue fleshy or cartilaginous. middle leg f. Feby 1852

Mantis (largest green sp) lower lip dissected with gt ease. the membranous tegument of body between the pieces easy examination. mentum transverse horny hydrate, when in life I saw the labium moved up & down, the membrane being the hinge, mentum remaining unmoved.

Jany. 1853

palpi
lobes
stem
membran
labium
mentum
cornous

tongue
labium

tongue is attached at base to inside of mentum

Steiradoma ?   No 10

Lamiidae
(broad flat maxilla see
yellow beneath
joint hooks)

Feby 1852

The paraglossa (lingua? Orthopt?) is a
deeply 2 cleft organ attached to inside of
labium, mentum is small triangular
& coriaceous, (flexible). the labi palpi are set
on joint of labium which is in one piece
(seated) with the mentum & not hidden behind
as in Pentamerous Coleop.   June 19 1853

Cyclica —

and magnified

mand.

ment. labi palpi
tongue behind

& dens &c laus ft.

Cythna (? No rough) mouth minute — men
very short escavate — lab. visible within thence
palpi — tongue with a few cilia behind — claw
thickened base? — & others, differing from out
— ant. 4–8 jts inner fore angle produced given
serrate appearance gradi enlarged forming
a club — max. single lobe oval — minute

Jany 1853

from obs. p. ⊗

Geodephaca — Trechidae
Bembidiidae

No 13

not seg.

mand.

carnivorous
further dissection

maxilla
correct

mentum
& labi palpi correct

Decem 1852

Genus Ega Laporte

Gen char. Head globulous, behind the eyes
annulate & fall with a much contracted
.... & no basiform in the prothr. Palpi
terminal joint large swollen clavate
... scarcely ... apex a small cla
... curved point. mentum with small
... semicircular notch. Mandibles
long tapering acute crossing each other
base edental fore edge slightly serrulate
... at the point of crossing of mand.
... ... base head, forwards globulous
... much narrowed behind, ... the waist
with the eyes again right as the waist
palpi ... with deep transverse concavity ...
... the head ... slender tarsal jts
... as long as 3 following in all 4 fore
... lobe, convex, slightly iridescent.

Ega globuliceps ... pubescent ... with a glossy bronzy tinge especially on head, elytra
each with 3 white spots 3 of small ...

Geodephaga Stenolophides

No 14

Stenorolobrus Dej ?
4 jts antle dilata scandate ... beneath
setose, and stiff hairs — max ... straight
no teeth at tip — dense fing of hairs
mandi broad stout ... within
Decem. 1852 — corrected tarsi Novm 19/54

Further Observation on Ega
The term. joint of palpi I quite
mistook on 1st dissection.
the small point or hook is
the apex of true 4 the terminal
jt in max. & 3rd in lab. —
I have no doubt Ega
is wrongly placed by Laporte
& belongs to Brachinidae
near Castorina. The lijula
& paraglossa I cannot observe
having only a single lens —

... serrate with inner angle
of upper margin of each jt being
produced acute. eyes much ...
forming 2 lobes (a) one smaller
root of ant. other large below it.
mand. (b) triangular short about
tip feebly bifid or ... tooth
in middle of inner edge.
Labrum 4 dentate front edge some
(c) fringed — narrow ... at 1/3
of front edge of clypeus. maxilla
(d) curved bill-shaped densely
fringed within — lobe rounded
clavate longer than stem, densely
fringed at tip ... edge — palpi
short, length of lobe, thick some
jts. term. 2 as long as any of other
truncate — ment. (e) fore edge in
.. each side angular broad not
f. labium base appearing beyond
tip of ment. lobes each of lijula
delivery narrow points base
fringed — g. ... within — h. lip
of fore lip globulous

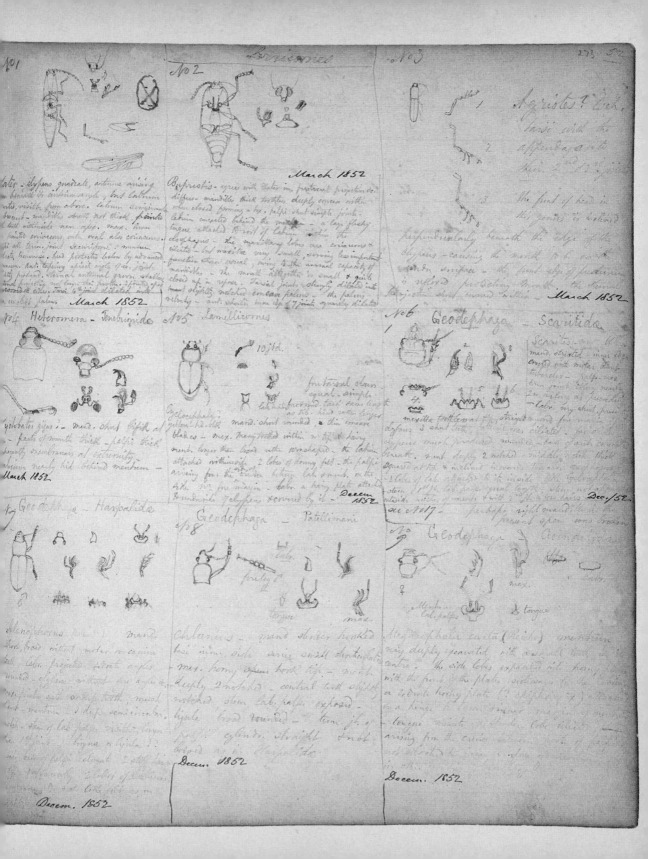

March 1852

later – clypeus quadrate, antennæ missing
in beneath its antinnae-eye, but labrum
into, visible from above. labrum semijoined
beneath – mandible small, not thick, pointed
of teeth withinside near apex. max. firm
pointed coriaceous, with oval also coriaceous.
the all terin joint securiform – mentum
hate transverse, but protected below by advanced
mum. ant tapering apical angle prom. joints
ento produced, sternal antennal groove, shallow
tend pointing, not time the finstin, if joints spi
rounded alter tarsi 4th joint dilated with
a visible palm. March 1852

Buprestis – agrees with later in pectoral projection, de-
differs – mandible thick toothes, deeply concave within
when closed forming a box. palpi short single joints.
labrum emarginate behind the mentum – a large fleshy
tongue attached to root of labrum within the
œsophagus – the maxillary lobes are coriaceous &
ciliate – but maxilla very small, serving less important
function than usual, owing to the unusual capacity of
mandibles – the mouth altogether is small & quite
closed up in repose. Tarsal joints obliquely dilated into
most slightly notched concave palms – the palms
velvety – ant. shorter than elp of 7 joints squarely dilated

Agriostes? lich
tarsi with the
appendage into
their 2nd & 3rd joint

3   the front of head in
this genus is inclined
perpendicularly beneath the edge of the
clypeus – causing the mouth to be seen with
greater surface – the front edge of produced –
is reflexed protecting mouth the teeth
projection short inward within – March 1852

No 4 Heteromera – Tenebrionidæ
No 5 Lamellicornis
No 6 Geodephaga – Scaritidæ

Cyclocephala gigas? – mand. short bifid at
– facets of mouth thick – palpi thick
seemingly membranous at extremity.
Antenna nearly hid behind mentum –
March 1852

10 std.

frontonsal claw
equal – simple
labrum, front & tarsi same length
as tib. hind rather larger

Cyclocephala?
yellow hd. blk. mand. short rounded – thin concave
blades – max. many toothed within w. tip & hairy
ment. longer than broad with unshaped – the labium
attached withinside – 2 lobes of horny felt – the palpi
arising from the centre between lab & ment. on the
side in free margin – labr. a horny plate attached
downwith & clypeus & covered by it – Decem 1852

Scaritidæ
mand. straight, inner edge
crested with molar teeth
tubercles – palpi–six
horn joint billow, narrow
the subocry as if inside
– labr. very short projecting –
maxilla toothless at tip, straight & inner face margin
clypeus 3 short inter. labral ones, ciliated – the
clypeus much produced, rounded – base faint, central
break. ment. deeply 2 notched – middles teeth thick
squared at tip & inclined inward. the as. of clypeus
3 lobes of lab attached to it inside of the tip of clypeus
ment. & palpi, lab. palps equal length – chin notched
withinside – with & with a few hairs – Dec./52
see No 17 – perhaps right mandible is the
present one was broken

No Geodephaga – Acinolinæ

No 7 Geodephaga – Harpalidæ
No 8 Geodephaga – Patellimani

Selenophorus ...? mand.
short, broad without molar or central
tell. labr. projected & teeth angles
rounded – clypeus without free angles –
membrane with only 2 teeth, much
bent – mentum 1 deep semicircular
notch. chin of lab palps visible, larger
... joined – ligula & ligula ?
... base of palpi inclined 2 with hairs
... diaphanously 2 lobes of diaphanous
membrane – & not like felt as in
... Decem. 1852

Chlænius – mand. slender hooked
base inner side series small denticulate
– max. horny apex hook tip – mouth
deeply 2 notched – central teeth oblong
notched chin lab. palps exposed –
ligula broad rounded – term. jt. of
palpi cylindr. straight & not
bowed as in Harpalidæ.
Decem 1852

Acinopella curta (Leach) mentum
very deeply separated with a small tooth in
centre. the side lobes expanded into horn
with the front of the plate, out wing to the lower
a toothed horny plate (? clypeus ? etc.) attached
by a hinge to lower inner margin of mentum
– tongue minute & slender lobe ciliate
arising from the centre – & maxilla root of palpi
not attached the lower ...
...
Decem. 1852

No 1—a dorsal
1c   1d

No 1 Scolopendra — ant. 17 joints — body 23 seg^{ts} including hd. & rest. of maxillary feet. feet 21 pairs excl. max. feet. The Myriapods shew the nature of the head & mouth appendages in the Articulata. 1c mandibles with small labrum. 1d the maxillæ & palpi at least the homologues of same in the Hexapoda; here resembling rather the labium & like it closing the mouth below. This shews that maxillæ & labium are of same nature & that the ~~legs~~ palpi are legs transformed. ⟨Mr L⟩ the maxillary feet No 2 are homologues of labial palpi in Hexapods but here do not serve in mastication have rather prehens raptorial functions. that they truly belong to head shewn by the fact that the seg^{ts} to wh. they belong has only a ventral not a dorsal plate. Thus we see that the head of Hexapods is composed of several seg^{ts} consolidated for here we see the hind most seg^t (the labial) in a transitionary state

Orthoptera — Locusta

membrane a corresposplat
membrane of throat
inner lobe   outer lobe
support
of palp moving when it moves

Labium of Acanthodis 40 Cwm tible
sp. allied to A. coronata. —
The palp arises from nr. the
base of the stem & has its support
separated by a deep furrow or
apparent suture from the stem
— the support at bottom arises from
the root hypocheil membrane & &
consists of a small lunar out-com-
cave plate, then a smaller lunar plate
above & exterior bit — this 2nd piece
is immediately articulated with the
basal jt. of palp & moves with it
in life. — The outer lobe "galea"
do exactly similar to gal. of the max.
is elbowed externally — it passes the
inner lobe & de top & reaches outside
to nr. the base of palp. But it is
attached throughout to the stem & has
no separate movement. C^y Nov/58

side view abd'. seg'.

7th
8th
9th
10th?

side view abdom' seg's

7th
8th
9th
10th?

♂

Proscopia — The ♂ & ♀ here
sketched, are perhaps not of
the same species. —— This
genus consists of species which
are not so vigourous leapers
as Acridia — they leap but
short distance & with little
force. This is borne out by the

structure of the hind legs, particularly the apex of tibiæ, which instead
of a cluster of curved teeth, have
only one or two small ones. — The
tarsi have only 3 entire joints.
but the basal one is divided
on the under surface into three, showing
that it is the union of 3 joints.
— The antennæ have 7 joints.
the apical one is elongate & has
indications of the union of 3 jts.
— The mouth offers nothing
essentially different from the
other Acrididæ. —— The
abdomen I think consists of
10 segments — 7 of which only
are entire with dorsal &
abdominal arches — the 3 terminal
are modified much from being

employed as accessory organs
in the sexual parts &c.
— If we compare the abdominal
seg's in this genus & in Libellula
we find confirmation of some
important laws in the morphology of the Articulated animals
— The Libellula have also 10
seg's. but as they are furnished
with powerful organs of
flight. the basal 2 seg's are
shortened & broadened closely
in connection with the
metathorax. the Proscopia
on the contrary. being apterous
have the basal 2 seg's fully
developed like the rest &
without signs of subserving
~~function~~ in the metathorax
in its functions. — But in
the Libellula the apical
seg's, at least the 8th & 9th
are ~~fully~~ entire — If in the
Proscopia they are much
shortened & modified to subserve
the sexual organs. In one of
our figures the apical
seg's of the ♂ are figured, in the
other those of ♀ — in the ♂ the
ventral arch of 8th forms. a large
horny convex plate above & the
9th a similar plate curved
upwards & pointed. In the ♀
The ventral arches of the 8th & 9th
seg's are each divided into a pair
of elongate, horny, tooth-shaped
processes.

Novem 6 - 1853 — altera do char
very wet day

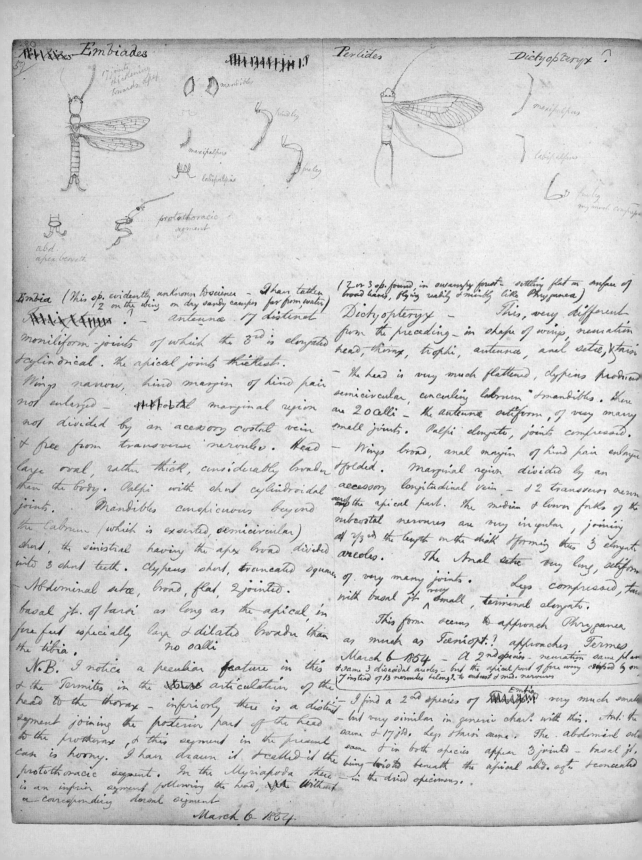

Embiades   Perlides   Dictyopteryx?

(3 joints thickening towards apex)

mandibles

maxipalpus

labipalpus

hindleg

foreleg

protothoracic segment

abd. apex beneath

maxipalpus

labipalpus

foreleg very much compressed

Embia (This sp. evidently unknown to science — I have taken ½ on the wing on dry sandy camps far from water)
— antennæ 17 distinct moniliform-joints of which the 3rd is elongated & cylindrical. the apical joints thickest.

Wings narrow, hind margin of hind pair not enlarged — total marginal region not divided by an accessory costal vein & free from transverse nervules. Head large oval, rather thick, considerably broader than the body. Palpi with short cylindroidal joints. Mandibles conspicuous beyond the Labrum (which is exserted, semicircular) short, the sinistral having the apex broad divided into 3 short teeth. Clypeus short, truncated square. — Abdominal setæ, broad, flat, 2 jointed. basal jt. of tarsi as long as the apical, in fore feet especially large & dilated broader than the tibia.   no calli

N.B. I notice a peculiar feature in this & the Termites in the articulation of the head to the thorax — inferiorly there is a distinct segment joining the posterior part of the head to the prothorax, & this segment in the present case is horny. I have drawn it & called it the protothoracic segment. In the Myriapoda there is an inferior segment following the head, without a corresponding dorsal segment.

March 6. 1854.

(2 or 3 sp. found in swampy forest — settling flat on surface of broad leaves, flying readily & nimbly like Phryganea)

Dictyopteryx — This, very different from the preceding — in shape of wing, neuration, head, thorax, trophi, antennæ, anal setæ, & tarsi — the head is very much flattened, clypeus produced semicircular, encircling labrum & mandibles. there are 2 ocelli — the antennæ setiform, of very many small joints. Palpi elongate, joints compressed. — Wings broad, anal margin of hind pair enlarged & folded. Marginal region divided by an accessory longitudinal vein — & 2 transverse nerven on the apical part. the median & lower forks of the subcostal nervures are very irregular, joining at ⅔ds the length on the disk & forming thus 3 elongate areoles. The Anal setæ very long, setiform of very many joints. Legs compressed, tarsi with basal jt. very small, terminal elongate.

This form seems to approach Phryganea as much as Tæniopt.? approaches Termes
March 6 1854 — A 2nd species — neuration same plan & same 3 discoidal areoles — but the apical part of fore wing crossed by only 7 instead of 13 nervules belong? to subcost & med. nervures

— I find a 2nd species of Embia very much smaller — but very similar in generic char.rs with this. Ant. the same & 17 jtd. Legs & tarsi same. The abdominal setæ same & in both species appear 3 jointed — basal jt. being two to beneath the apical abd. sgt. & concealed in the dried specimens.

sinistral   dextral

semi
corneous

membrane   corneous

labrum
seen beneath

Megacephala Klugii ♀
— the mentum & labium are
a curious study. the
mentum is not all in one
plane like that of the Carabi
the lateral lobes fall inward
& encircle the labium behind
on the side of the œsophagus
the labium is thus reduced
to a small corneous point
with 2 bristles & a pencil of
hairs on ea. side, situated
together with roots of palpi
in the space between the
corneous circuit of mentum behind
& the central tooth of same
in front. Ega 15 aug't. 1856
M. Klugii now very abundant on
the beach, running in moonlight.

Singular species of Lampyridæ
forming I think not only n. g, but
new tribe of the Lampyrides
— It was luminous in three
parts of the body 1st the mouth
which formed a brilliant nucleus
of phosphorescent light — 2 the
nucleus, the whole of the neck
was luminous but the nucleus
was on the nucleus — 3rd the
3 apical segts of abdomen on
each side of each of which was
a nucleus of light. ——
caught by sweeping in wood
near a rotting log — Ega
August 1856 ——

no1

Passalus –
hard – same substance as
integuments – 6 Sept/56

ligula very
hard – same substance as
integuments – 6 Sept/56

Dacne
Erotylides
Engidiformes
rare at
Ega on
boleti
but generally found cast on shore of
lake after a storm at night

Staphylinus 1 X Er

Staphylinus segmentarius Er
– the notch in the ligula is so
very minute as scarcely to be
perceptible with powerful glass.
– paraglossae much longer than lig.
– Insect like a Philonthus or
rather a Belonuchus, not at
all like typical sp. of Staphylinus
– common in dung at Ega – also
amongst putrid mandiocca parings
Ega 30 Octo 1856

7 lines

Staphylinus no 783 (blue) undescr. in B.
Ega common . putrid bark , fruit &c

fore leg ♂
3 jts with squammæ

mentum

leg & paragl.

mand & labrum

Pericallide — beautiful insect
found running over old stumps, amongst
Boleti. very flat in form (Lin in colours
　　　　　(smooth elytra &c)
the labrum is ample & strongly
rounded in front. The mandibles
are short & very broad & flat, the
dextral having a tooth in middle
interior edge — the muzzle is short
this charr. together with mandibles
distinguishes them from Coptodera
(sup. g. no 38.) ——— The mentum
is without tooth in notch. the
paraglossa are attached to
ligula up to front edge from
the angle of which they diverge
& are not much longer.
the fore tarsi of ♂ has 3 jts
squammæ like Coptodera
some Lebia &c. the 4th jt
of all is quite simple & the
claws are not much curved
& have only 4 or 5 short pectination
for basal ½ of inner edge &
the claw not being thickened
in that part. — all the
palpi have their term. jts
cylindrical. otherwise.
Ega 9 Sept. 1856.

Coptodera? reddish ferruginous
with black fasciæ as above. —
found running amongst boleti in
comp. with no 1 —
Mandibles, muzzle & labrum
together rather more elongated than
in no 1 — labrum notched in mid
front edge. — Mentum as in
no 1. Paraglossæ attached
to ligula exactly as no 1, but
not surpassing it in length, or
by an extremely short amount.
legs, tarsi & claws exactly
as no 1 — the ♂ having
3 jts of tarsi squammated as
in all their allied species.
Ega 9 Sept. 1856

no 3

labrum
leg & paragl.

feet of
no 1 & 2

dirty green bronze
some coppery
legs, ant & underbody
piceous

a species not uncommon amongst
branches & chinks of newly felled trees
— muzzle much elongated & flattened
— the insect is excessively lik. Agonum
parvimpunctatum in facies & colours.
Ega 18 Sept. 1856

no 4

ferrugin
red
bright
green brassy
legs, exterior & underbody, reddish

another species in c.t with no 3 — about
newly felled timber especially where sap exudes

Curious Carabide
all black, shape of Anchomeni
in fact very much like A. angus-
ticollis of Europe — but elytra
obliquely sinuæ-truncate like
the Coptoderæ. — found in
company with no 1 & 2 running
amongst Boleti.
the mentum has a strong tooth.
the paraglossæ are attached
to ligula exactly as in 1 &
surpass it in length exactly in
same way. The labrum
is like no 2. The dextral
mandible has a more
prominent tooth than no 2
the fore tarsi of ♂ have
shorter jts, but are furnished
with squammæ in same way.
the claws are of a decidedly
different type being simple
like those & strongly arched
like those of Cicindela
& the other cursorial
Carabidi. I think it
comes near Thyreopterus
Ega 9 Sept. 1856

see below CPS

Omoplata dicha

Liachnophorus

ligula & paraglossa
the first a. is membranous
rather thinner than
texture of paraglossa

ligula & paraglossa
of a genus of Harpalida
allied to Bradycellus
the ♂ having ½ of
five tarsi only dilated
& furnished with squamma

Species of sp. ?
Bradytus —
ferruginous sp.
on fuscous. Glossy
common muddy
margins of streams

young
Larva — head
exserted i.e. not covered
by protho. hds. pro-
blane corneous —
the rest of abd. body
soft pale greenish yell.
the protho. & ex. add.

sect. on ea. side below the spiracle have a sharp spiniform
tubercle rather elongate. apical seg. has a pair of
vertical or oblique recurved spines which carry the
excrements in life. — Mother tend the young larva
like the Pentatomidae do their own. Lit is very difficult
to drive it off except by force.

contain?
jts. with
recurved squama

paraglossae &
ligula
wholly membranous

Pupa
of single
Dichra

remains of skin
of larva which
envelopes the pupa
beneath

Larva of Eumorphus
(the common brassy sp. with
pale margins) rather
actively moving over the
dead branches — nibbling
the minute lichens on the
bark; in co with imago. —
the antennae are long &
formed apparently of 3 jts. but I think they
articulate only by their base. the basal rounded joint
being small & pale in colour, the 2nd very small &
separated from the main piece of antennae probably by
a constriction only. There are 4 ocelli
on each side of the hd. 3 in triangle above & one
below the antennae. the mandibles are acute
outline as fig — maxillae & palps as fig. the
legs have one tarsal jt & a claw. the
body has only 11 dorsal segts. the substance
is fleshy. the blk. dorsal parts granulated, & rather
harder in texture. the margins of the body are
fulvo-ochraceous the rest of dorsal surface
fuliginous-blk. speckled with round pale
scales —   Ega 28 Sept 1858

max.
ciliated
on outside
at tip
not toothed

blk. spines rufous

Larva of Ægithus Surinamensis?
See also fig No 822 blue
Each segment of body from the meso tho.
to the penultimate with a row of 4
long spiniferous corneous tubercles (a)
the prothorax having 2 rows of the same &
the anal segt. 2 longer & slender pointed ones.
head with moderately short rigid cylindrical
antennae consisting of 3 jts (b) mandibles short
broad curved apex truncate with 3 equal teeth
(c) ocelli apparently 6 in 2 rows behind each
the e — Maxill. one lobe obtuse with a
few minute denticuli at tip — palps short
conical with 3(?) obscure jt. Labipalps
2 arti. claw with a minute tooth at base
— feeds on hard bolete on old palings

remains
of larva-spines

Gyrinidæ - Gyretes probably - I
cannot understand the character
of the family given by Lac. as
far as concerns the antennæ
which as terminal jt very large
composing the preceding 7th united
I find them as in the above fig.

larva of ~~Petho~~ Tauroma
smaragdina Boh.   Sparingly on
me tree in co: with perfect insect - Ega -
the body is fleshy broad, oblong, its
margin projecting considerably beyond
the legs in the prothx. deeply
emarginated in front - head being
hidden from above. - the sides of
thorax & abdomen having a
fringe of fleshy processes - 8 on
each side of thorx - & one to each of
abdl. segts. the terminal segt. having
a pair of very long diverging processes
curved inwards at end. - of a duller
consistence than the lateral processes
is dark brown polished with an
margin & fringe are testaceous - Ega 23 June /58

Larva & pupa of Dolichosoma Batesii. The larva has
4 spines on each side of prothx. 3 to mesothx & 2 to metathx
one on each side 1st to 8th seg. abd., those of 7th & 8th being
twice the length of the others. The terminal seg. is prolonged into
a strong horny piece furnished with 2 very strong spines
one each side, the organ being erectile in life over the
back, all the spines except the 3 first on each side
are more or less denticulate.   The spines & a spot
on sides of body near their roots, which becomes gradually
more elongate towards tip of abdom., an 2 large spots
on pronotum & interrupted belt on metathorax
& a pair of small spots in dorsal surface of the
basal abdominal segts. blk. the rest of the
surface brilliant vermillion.   one claw to ea.
tarsus -   antennæ minute - 2 jts - a thin one
with a broader one serving as base. Mandibles short
thick concave within. Labrum sinuated in middle
- Max. palpi - apparently of same structure as antennæ
& same no. of jts.   There is an appearance as of
a cluster of small ocelli on each side of hd. above
& outside of antennæ.   The edge of prothx. covers
the head above, but does not project much beyond
it.

In the pupa the thoracic segts. have no lateral
processes. The abdominal have one on each side
a thin depressed simple blade - gradually decreasing
in size from the 1st abd. segt. downwards
Ega 16 June /58

- although of same colour. The colour of body above
an oval spot in centre light fulvous testaceous - the

Buprest. (lab.d max. shd.)

Chrysothea tripunctatus Ld.

The max. have the 1st jt. twisted
the trochanter-like oscalating piece
(q. Blatta p. 45) is decidedly absent
The stem or 2nd jt. a. is an
irregularly quadrate corneous piece
whose external edge alone is of even
metallic hard substance as the rest
of the integuments, the inner parts
are corneous. The inner lobe
is a narrow stripe of semicorneous
membrane attached entirely to the
inner upper edge of _a_ & not articulated
— the outer lobe c is a thick
membranous organ apparently
composed of 2 equal sized joints
it is partially semicorneous & very
flexible. —

The mentum is very short &
broad — the upper part has a
semicircular membrane quite
free from corneous substance —
The supports of labial palpi
I consider are the remains of
the stems of labium (a, Blatta 45)
the lobes of labrium are entirely
wanting as in all Coleoptera —
b. I consider to h the ligula
corresponding with the ligula or tongue
h. of orthoptera. In Coleop. in
consequence of the collapse of the
labium the labial palpi are attached
to base of ligula. Ega 28 Septem /58

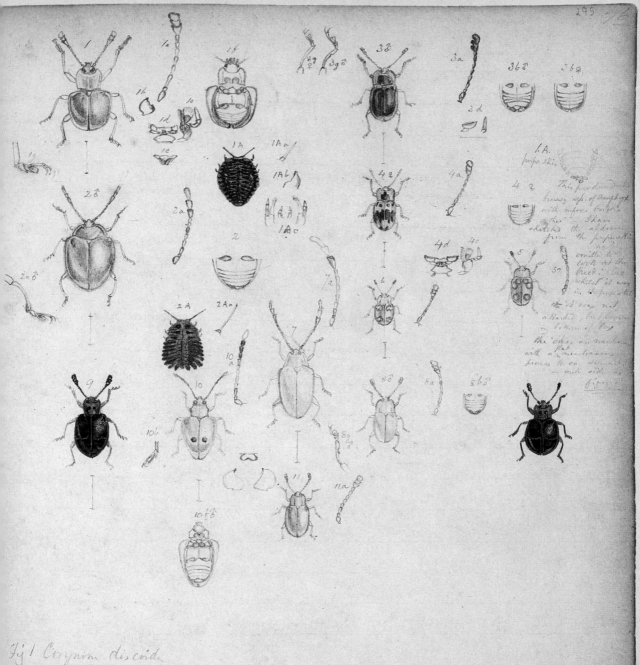

Fig 1 Corynum discoide
  1a antenna
  1b Mand.
  1c Max & palp
  1d Lab. palp
Fig 2 Stenotarsus obtusus

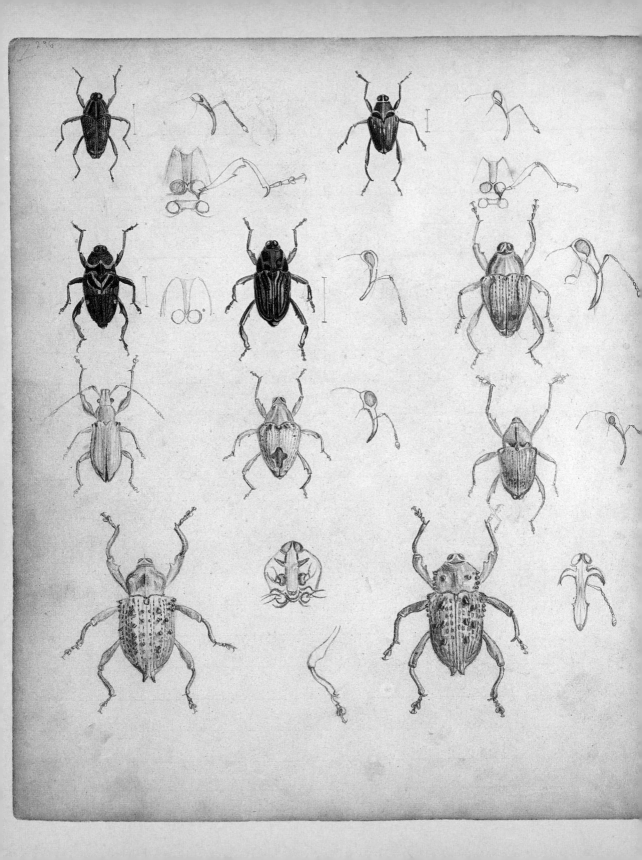

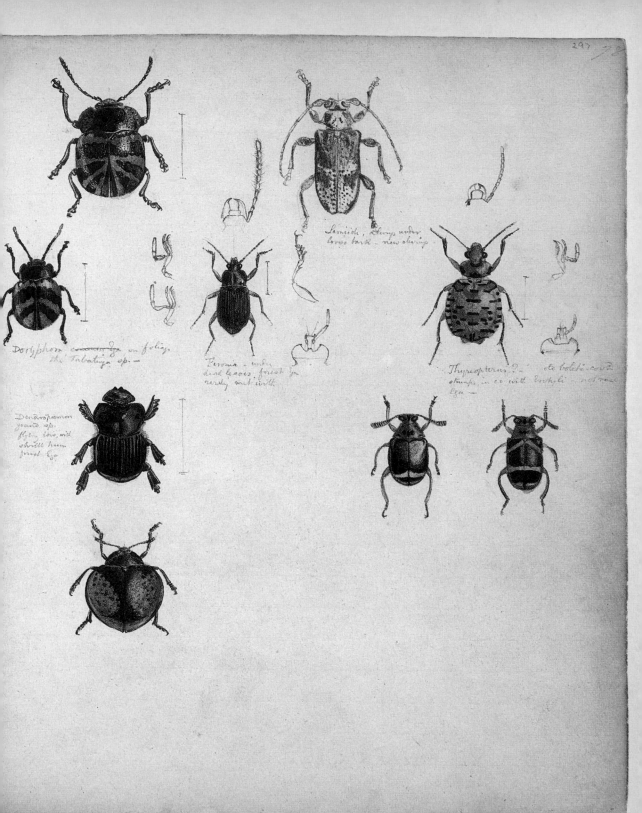

Doryphora crocandega on foliage
the Tabatiga sp.—

Peronia — under
dead leaves forest —
rarely met with

Lamicoli, always under
loose bark — new always

Thyreopterus ?— ete boleti-cora
stumps, in co with Trechli — not rare
lga —

Dendropremon
grand sp.
flies, low, will
shrill hum
forest Lga

The Naturalist on the
River Amazons

SANTARÉM to EGA

THE RIO NEGRO broadens considerably from its mouth upwards, and presents the appearance of a great lake; its black-dyed waters having no current, and seeming to be dammed up by the impetuous flow of the yellow, turbid Solimoens, which here belches forth a continuous line of uprooted trees and patches of grass, and forms a striking contrast with its tributary. In crossing, we passed the line, a little more than half-way over, where the waters of the two rivers meet and are sharply demarcated from each other. On reaching the opposite shore, we found a remarkable change. All our insect pests had disappeared, as if by magic, even from the hold of the canoe: the turmoil of an agitated, swiftly flowing river, and its torn, perpendicular, earthy banks, had given place to tranquil water and a coast indented with snug little bays, fringed with sloping sandy beaches. The low shore and vivid light green endlessly varied foliage, which prevailed on the south side of the Amazons, were exchanged for a hilly country, clothed with a sombre, rounded, and monotonous forest. Our tedious voyage now approached its termination; a light wind carried us gently along the coast to the city of Barra, which lies about seven or eight miles within the mouth of the river.

Santarém

About the month of April, when the water rises to [the extent that it overflows the campo], the trees are covered with blossom, and a handsome orchid, an *Epidendron* with large white flowers, which clothes thickly the trunks, is profusely in bloom. Several kinds of kingfisher resort to the place: four species may be seen within a small space: the largest as big as a crow, of a mottled-grey hue, and with an enormous beak; the smallest not larger than a sparrow. The large one makes its nest in clay cliffs, three or four miles distant from this place. None of the kingfishers are so brilliant in colour as our English species. The blossoms on the trees attract two or three species of humming-birds, the most conspicuous of which is a large swallow-tailed kind (*Eupetomena macroura*), with a brilliant livery of emerald green and steel blue. I noticed that it did not remain so long poised in the air before the flowers as the other smaller species; it perched more frequently, and sometimes darted after small insects on the wing. Emerging from the grove there is a long stretch of sandy beach; the land is high and rocky, and the belt of wood which skirts the river banks is much broader than it is elsewhere. At length, after rounding a projecting bluff, the bay of Mapirí is reached. The river view is characteristic of the Tapajós: the shores are wooded, and on the opposite side is a line of clay cliffs, with hills in the background clothed with rolling forest. A long spit of sand extends into mid-river, beyond which is an immense expanse of dark water, the further shore of the Tapajós being barely visible as a thin grey line of trees on the horizon. The transparency of air and water in the dry season when the brisk east wind is blowing, and the sharpness of outline of hills, woods, and sandy beaches, give a great charm to this spot.

The little pools along the beach were tenanted by several species of fresh-water molluscs. The most abundant was a long turret-shaped *Melania*, which swarmed in them in the same way as *Limnaeae* do in ponds at home. I found no *Limnaea*, nor indeed any European genus of fresh-water mollusc, in the Amazons region. After the first storms of February the coast is strewn with large apple-shells (*Ampullaria*). They are not inhabitants of the pools on this side of the river, but are involuntary visitors, being driven across

by the wind and waves with masses of marsh plants from the low land of the opposite shore. A great many are dead shells, and more or less worn. In showery weather I seldom came this way without seeing one or more water snakes of the genus *Helicops*. They were generally concealed under the heaps of thick aquatic grasses cast ashore by storms; and when exposed, always made off straight for the water. They glided along with such agility that I rarely succeeded in capturing one, and on reaching the river they sought at once the bottom in the deepest parts. I believe these snakes are swept from the marshy land of the western shore with the patches of grass and the *Ampullariae* just mentioned. Other reptiles and a great number of insects are blown or floated over in the same way by the violent squalls which occur in January or February. None of the species take root on the Santarém side of the river. Sometimes myriads of coleopterous insects, belonging to about half a dozen kinds, are blown across, and become perfect pests to the town's people for two or three nights. swarming about the lights in every chamber. They get under one's clothing, or down one's back, and pass from the oil-lamp on to the furniture, books, and papers, smearing everything they touch. The open shops facing the beach become filled with them, and customers have to make a dash in and out through the showers that fall about the large brass lamps over the counter, when they want to make a purchase. The species are certainly not indigenous to the eastern side of the river; the hosts soon disappear; those which cannot get back must perish helplessly, for the soil, vegetation, and climate of the Santarém side are ill suited to the inhabitants of the opposite shore.

INSTINCT OF LOCALITY

Whilst resting in the shade during the great heat of the early hours of afternoon, I used to find amusement in watching the proceedings of the sand-wasps. A small pale green kind of *Bembex* (*Bembex ciliata*), was plentiful near the bay of Mapirí. When they are at work, a number of little jets of sand are seen shooting over the surface of the sloping bank. The little miners excavate with their fore feet, which are strongly built and furnished with a fringe of stiff bristles; they work with wonderful rapidity, and the

sand thrown out beneath their bodies issues in continuous streams. They are solitary wasps, each female working on her own account. After making a gallery two or three inches in length in a slanting direction from the surface, the owner backs out and takes a few turns round the orifice apparently to see whether it is well made, but in reality, I believe, to take note of the locality, that she may find it again. This done, the busy workwoman flies away; but returns, after an absence varying in different cases from a few minutes to an hour or more, with a fly in her grasp, with which she re-enters her mine. On again emerging, the entrance is carefully closed with sand. During this interval she has laid an egg on the body of the fly which she had previously benumbed with her sting, and which is to serve as food for the soft, footless grub soon to be hatched from the egg. From what I could make out, the *Bembex* makes a fresh excavation for every egg to be deposited; at least in two or three of the galleries which I opened there was only one fly enclosed.

I have said that the *Bembex* on leaving her mine took note of the locality: this seemed to be the explanation of the short delay previous to her taking flight; on rising in the air also the insects generally flew round over the place before making straight off. Another nearly allied but much larger species, the *Monedula signata*, whose habits I observed on the banks of the Upper Amazons, sometimes excavates its mine solitarily on sand-banks recently laid bare in the middle of the river, and closes the orifice before going in search of prey. In these cases the insect has to make a journey of at least half a mile to procure the kind of fly, the Motúca (*Hadaus lepidotus*), with which it provisions its cell. I often noticed it to take a few turns in the air round the place before starting; on its return it made without hesitation straight for the closed mouth of the mine. I was convinced that the insects noted the bearings of their nests and the direction they took in flying from them. The proceeding in this and similar cases (I have read of something analogous having been noticed in hive bees) seems to be a mental act of the same nature as that which takes place in ourselves when recognising a locality. The senses, however, must be immeasurably more keen and the mental operation much more certain in them than it is in man; for to my eye there was absolutely no landmark on the even surface of sand which

could serve as guide, and the borders of the forest were not nearer than half a mile. The action of the wasp would be said to be instinctive; but it seems plain that the instinct is no mysterious and unintelligible agent, but a mental process in each individual, differing from the same in man only by its unerring certainty. The mind of the insect appears to be so constituted that the impression of external objects or the want felt, causes it to act with a precision which seems to us like that of a machine constructed to move in a certain given way. I have noticed in Indian boys a sense of locality almost as keen as that possessed by the sand-wasp. An old Portuguese and myself, accompanied by a young lad about ten years of age, were once lost in the forest in a most solitary place on the banks of the main river. Our case seemed hopeless, and it did not, for some time occur to us to consult our little companion, who had been playing with his bow and arrow all the way whilst we were hunting, apparently taking no note of the route. When asked, however, he pointed out, in a moment, the right direction of our canoe. He could not explain how he knew; I believe he had noted the course we had taken almost unconsciously: the sense of locality in his case seemed instinctive.

The *Monedula signata* is a good friend to travellers in those parts of the Amazons which are infested with the blood-thirsty Motúca. I first noticed its habit of preying on this fly one day when we landed to make our fire and dine on the borders of the forest adjoining a sand-bank. The insect is as large as a hornet, and has a most waspish appearance. I was rather startled when one out of the flock which was hovering about us flew straight at my face: it had espied a Motúca on my neck and was thus pouncing upon it. It seizes the fly not with its mandibles but with its fore and middle feet, and carries it off tightly held to its breast. Wherever the traveller lands on the Upper Amazons in the neighbourhood of a sand-bank he is sure to be attended by one or more of these useful vermin-killers.

Mason Wasps

In the lower part of the Mahicá woods, towards the river, there is a bed of stiff white clay, which supplies the people of Santarém with material

for the manufacture of coarse pottery and cooking utensils: all the kettles, saucepans, mandioca ovens, coffee-pots, washing-vessels, and so forth, of the poorer classes throughout the country, are made of this same plastic clay, which occurs at short intervals over the whole surface of the Amazons valley, from the neighbourhood of Pará to within the Peruvian borders, and forms part of the great Tabatinga marl deposit. To enable the vessels to stand the fire, the bark of a certain tree, called Caraipe, is burnt and mixed with the clay, which gives tenacity to the ware. Caraipe is an article of commerce, being sold, packed in baskets, at the shops in most of the towns. The shallow pits, excavated in the marly soil at Mahicá, were very attractive to many kinds of mason bees and wasps, who make use of the clay to build their nests with. I spent many an hour, watching their proceedings: a short account of the habits of some of these busy creatures may be interesting.

The most conspicuous was a large yellow and black wasp, with a remarkably long and narrow waist, the *Pelopseus fistularis*. It collected the clay in little round pellets, which it carried off, after rolling them into a convenient shape in its mandibles. It came straight to the pit with a loud hum, and, on alighting, lost not a moment in beginning to work; finishing the kneading of its little load in two or three minutes. The nest of this species is shaped like a pouch, two inches in length, and is attached to a branch or other projecting object. One of these restless artificers once began to build on the handle of a chest in the cabin of my canoe, when we were stationary at a place for several days. It was so intent on its work that it allowed me to inspect the movements of its mouth with a lens whilst it was laying on the mortar. Every fresh pellet was brought in with a triumphant song, which changed to a cheerful busy hum when it alighted and began to work. The little ball of moist clay was laid on the edge of the cell, and then spread out around the circular rim by means of the lower lip guided by the mandibles. The insect placed itself astride over the rim to work, and, on finishing each addition to the structure, took a turn round, patting the sides with its feet inside and out before flying off to gather a fresh pellet. It worked only in sunny weather, and the previous layer was sometimes not

quite dry when the new coating was added. The whole structure takes about a week to complete.

Mason Bees

But the most numerous and interesting of the clay-artificers are the workers of a species of social bee, the *Melipona fasciculata*. The *Meliponae* in tropical America take the place of the true *Apides*, to which the European hive-bee belongs, and which are here unknown; they are generally much smaller insects than the hive-bees and have no sting. The *M. fasciculata* is about a third shorter than the *Apis mellifica*: its colonies are composed of an immense number of individuals; the workers are generally seen collecting pollen in the same way as other bees, but great numbers are employed gathering clay. The rapidity and precision of their movements whilst thus engaged are wonderful. They first scrape the clay with their mandibles; the small portions gathered are then cleared by the anterior paws and passed to the second pair of feet, which, in their turn, convey them to the large foliated expansions of the hind shanks which are adapted normally in bees, as everyone knows, for the collection of pollen. The middle feet pat the growing pellets of mortar on the hind legs to keep them in a compact shape as the particles are successively added. The little hodsmen soon have as much as they can carry, and they then fly off. I was for some time puzzled to know what the bees did with the clay; but I had afterwards plenty of

Melipona bees gathering clay

opportunity for ascertaining. They construct their combs in any suitable crevice in trunks of trees or perpendicular banks, and the clay is required to build up a wall so as to close the gap, with the exception of a small orifice for their own entrance and exit. Most kinds of *Meliponae* are in this way masons as well as workers in wax and pollen-gatherers. One little species (undescribed) not more than two lines long, builds a neat tubular gallery of clay, kneaded with some viscid substance outside the entrance to its hive, besides blocking up the crevice in the tree within which it is situated. The mouth of the tube is trumpet-shaped, and at the entrance a number of the pigmy bees are always stationed apparently acting as sentinels.

Rest in the Forest

My best hunting ground was a part of the valley sheltered on one side by a steep hill whose declivity, like the swampy valley beneath, was clothed with magnificent forest. We used to make our halt in a small cleared place, tolerably free from ants and close to the water. Here we assembled after our toilsome morning's hunt in different directions through the woods, took our well-earned meal on the ground — two broad leaves of the wild banana serving us for a tablecloth — and rested for a couple of hours during the great heat of the afternoon. The diversity of animal productions was as wonderful as that of the vegetable forms in this rich locality. I find by my register that it was not unusual to meet with thirty or forty new species of conspicuous insects during a day's search, even after I had made a great number of trips to the same spot. It was pleasant to lie down during the hottest part of the day, when my people lay asleep, and watch the movements of animals. Sometimes a troop of Anús (*Crotophaga*), a glossy black-plumaged bird, which lives in small societies in grassy places, would come in from the campos, one by one, calling to each other as they moved from tree to tree. Or a Toucan (*Rhamphastos ariel*) silently hopped or ran along and up the branches, peeping into chinks and crevices. Notes of solitary birds resounded from a distance through the wilderness. Occasionally a sulky Trogon would be seen, with its brilliant green back and rose-coloured breast, perched for an hour without moving on a low branch. A number of large, fat lizards

The Jacuarú (*Teius teguexim*)

two feet long, of a kind called by the natives Jacuarú (*Teius teguexim*) were always observed in the still hours of mid-day scampering with great clatter over the dead leaves, apparently in chase of each other. The fat of this bulky lizard is much prized by the natives, who apply it as a poultice to draw palm spines or even grains of shot from the flesh. Other lizards of repulsive aspect, about three feet in length when full grown, splashed about and swam in the water; sometimes emerging to crawl into hollow trees on the banks of the stream, where I once found a female and a nest of eggs. The lazy flapping flight of large blue and black *Morpho* butterflies high in the air, the hum of insects, and many inanimate sounds, contributed their share to the total impression this strange solitude produced. Heavy fruits from the crowns of trees which were mingled together at a giddy height overhead, fell now and then with a startling "plop" into the water. The breeze, not felt below, stirred in the topmost branches, setting the twisted and looped Sipós in motion, which creaked and groaned in a great variety of notes. To these noises were added the monotonous ripple of the brook, which had its little cascade at every score or two yards of its course.

White Ants (Termites)

Whenever a colony of Termites is disturbed, the workers are at first the only members of the community seen; these quickly disappear through the

endless ramified galleries of which a Termitarium is composed, and soldiers make their appearance. The observations of Smeathman on the soldiers of a species inhabiting tropical Africa are often quoted in books on Natural History, and give a very good idea of their habits. I was always amused at the pugnacity displayed, when, in making a hole in the earthy cemented archway of their covered roads, a host of these little fellows mounted the breach to cover the retreat of the workers. The edges of the rupture bristled with their armed heads as the courageous warriors ranged themselves in compact line around them. They attacked fiercely any intruding object, and as fast as their front ranks were destroyed, others filled up their places. When the jaws closed in the flesh, they suffered themselves to be torn in pieces rather than loosen their hold. It might be said that this instinct is rather a cause of their ruin than a protection when a colony is attacked by the well-known enemy of Termites, the ant-bear; but it is the soldiers only which attach themselves to the long worm-like tongue of this animal, and the workers, on whom the prosperity of the young brood immediately depends, are left for the most part unharmed. I always found, on thrusting my finger into a mixed crowd of Termites, that the soldiers only fastened upon it. Thus the fighting caste do in the end serve to protect the species by sacrificing themselves for its good.

A family of Termites consists of workers as the majority, of soldiers, and of the King and Queen. These are the constant occupants of a completed Termitarium. The royal couple are the father and mother of the colony, and are always kept together closely guarded by a detachment of workers in a large chamber in the very heart of the hive, surrounded by much stronger walls than the other cells. They are wingless and both immensely larger than the workers and soldiers. The Queen, when in her chamber, is always found in a gravid condition, her abdomen enormously distended with eggs, which, as fast as they come forth, are conveyed by a relay of workers in their mouths from the royal chamber to the minor cells dispersed throughout the hive. The other members of a Termes family are the winged individuals: these make their appearance only at a certain time of the year, generally in the beginning of the rainy season. It has puzzled naturalists to make out

the relationship between the winged Termites and the wingless King and Queen. It has also generally been thought that the soldiers and workers are the larvae of the others; an excusable mistake, seeing that they much resemble larvae. I satisfied myself, after studying the habits of these insects daily for several months, that the winged Termites were males and females in about equal numbers, and that some of them, after shedding their wings and pairing, became Kings and Queens of new colonies; also, that the soldiers and workers were individuals which had arrived at their full growth without passing through the same stages as their fertile brothers and sisters....

A few weeks before the exodus of the winged males and females a completed Termitarium contains Termites of all castes and in all stages of development. On close examination I found the young of each of the four orders of individuals crowded together, and apparently feeding in the same cells. The full-grown workers showed the greatest attention to the young larvae, carrying them in their mouths along the galleries from one cell to another, but they took no notice of the full-grown ones. It was not possible to distinguish the larvae of the four classes when extremely young, but at an advanced stage it was easy to see which were to become males and females, and which workers and soldiers. The workers have the same form throughout, the soldiers showed in their later stages of growth the large head and cephalic processes, but much less developed than in the adult state. The males and females were distinguishable by the possession of rudimentary wings and eyes, which increased in size after three successive changes of skin.

Thus I think I made out that the soldier and worker castes are, like the males and females, distinct from the egg; they are not made so by a difference of food or treatment during their earlier stages, and they never become winged insects. The workers and soldiers feed on decayed wood and other vegetable substances; I could not clearly ascertain what the young fed upon, but they are seen of all sizes, larvae and pupae, huddled together in the same cells, with their heads converging towards the bottom, and I thought I sometimes detected the workers discharging a liquid from their mouths

into the cells. The growth of the young family is very rapid, and seems to be completed within the year: the greatest event of Termite life then takes place, namely, the coming of age of the winged males and females, and their exit from the hive.

It is curious to watch a Termitarium when this exodus is taking place. The workers are set in the greatest activity, as if they were aware that the very existence of their species depended on the successful emigration and marriages of their brothers and sisters. They clear the way for their bulky but fragile bodies, and bite holes through the outer walls for their escape. The exodus is not completed in one day, but continues until all the males and females have emerged from their pupa integuments, and flown away. It takes place on moist, close evenings, or on cloudy mornings: they are much attracted by the lights in houses, and fly by myriads into chambers, filling the air with a loud rustling noise, and often falling in such numbers that they extinguish the lamps. Almost as soon as they touch ground they wriggle off their wings, to aid which operation there is a special provision in the structure of the organs, a seam running across near their roots and dividing the horny nervures. To prove that this singular mutilation was voluntary, on the part of the insects, I repeatedly tried to detach the wings by force, but could never succeed whilst they were fresh, for they always tore out by the roots. Few escape the innumerable enemies which are on the alert at these times to devour them; ants, spiders, lizards, toads, bats, and goat-suckers. The waste of life is astonishing. The few that do survive pair and become the kings and queens of new colonies.

Voyage up the Tapajós

I was obliged, this time [for my excursion up the Tapajós], to travel in a vessel of my own; partly because trading canoes large enough to accommodate a Naturalist very seldom pass between Santarém and the thinly peopled settlements on the river, and partly because I wished to explore districts at my ease, far out of the ordinary track of traders. I soon found a suitable canoe; a two-masted cuberta, of about six tons' burthen, strongly built of Itaúba or stone-wood, a timber of which all the best vessels in the Amazons

country are constructed, and said to be more durable than teak. This I hired of a merchant at the cheap rate of 500 reis, or about one shilling and twopence per day. I fitted up the cabin, which, as usual in canoes of this class, was a square structure with its floor above the water-line, as my sleeping and working apartment. My chests, filled with store-boxes and trays for specimens, were arranged on each side, and above them were shelves and pegs to hold my little stock of useful books, guns, and game bags, boards and materials for skinning and preserving animals, botanical press and papers, drying cages for insects and birds, and so forth. A rush mat was spread on the floor, and my rolled-up hammock, to be used only when sleeping ashore, served for a pillow. The arched covering over the hold in the fore part of the vessel contained, besides a sleeping place for the crew, my heavy chests, stock of salt provisions and groceries, and an assortment of goods wherewith to pay my way amongst the half-civilised or savage inhabitants of the interior. The goods consisted of cashaga, powder and shot, a few pieces of coarse checked-cotton cloth and prints, fish-hooks, axes, large knives, harpoons, arrow-heads, looking-glasses, beads, and other small wares. José and myself were busy for many days arranging these matters. We had to salt the meat and grind a supply of coffee ourselves. Cooking utensils, crockery, water-jars, a set of useful carpenter's tools, and many other things had to be provided. We put all the groceries and other perishable articles in tin canisters and boxes, having found that this was the only way of preserving them from damp and insects in this climate. When all was done, our canoe looked like a little floating workshop.

FOREST OF AVEYROS

Aveyros may be called the head-quarters of the fire-ant, which might be fittingly termed the scourge of this fine river. The Tapajós is nearly free from the insect pests of other parts, mosquitoes, sand-flies, Motúcas and Piúms; but the formiga de fogo is perhaps a greater plague than all the others put together. It is found only on sandy soils in open places, and seems to thrive most in the neighbourhood of houses and weedy villages, such as Aveyros: it does not occur at all in the shades of the forest. I noticed it in most

places on the banks of the Amazons, but the species is not very common on the main river, and its presence is there scarcely noticed, because it does not attack man, and the sting is not so virulent as it is in the same species on the banks of the Tapajós. Aveyros was deserted a few years before my visit on account of this little tormentor, and the inhabitants had only recently returned to their houses, thinking its numbers had decreased. It is a small species, of a shining reddish colour, not greatly differing from the common red stinging ant of our own country (*Myrmica rubra*), except that the pain and irritation caused by its sting are much greater. The soil of the whole village is undermined by it: the ground is perforated with the entrances to their subterranean galleries, and a little sandy dome occurs here and there, where the insects bring their young to receive warmth near the surface. The houses are overrun with them; they dispute every fragment of food with the inhabitants, and destroy clothing for the sake of the starch. All eatables are obliged to be suspended in baskets from the rafters, and the cords well soaked with copaüba balsam, which is the only means known of preventing them from climbing. They seem to attack persons out of sheer malice: if we stood for a few moments in the street, even at a distance from their nests, we were sure to be overrun and severely punished, for the moment an ant touched the flesh, he secured himself with his jaws, doubled in his tail, and stung with all his might. When we were seated on chairs in the evenings in front of the house to enjoy a chat with our neighbours, we had stools to support our feet, the legs of which as well as those of the chairs, were well anointed with the balsam. The cords of hammocks are obliged to be smeared in the same way to prevent the ants from paying sleepers a visit.

The Anaconda

We had an unwelcome visitor whilst at anchor in the port of João Malagueita. I was awoke a little after midnight as I lay in my little cabin by a heavy blow struck at the sides of the canoe close to my head, which was succeeded by the sound of a weighty body plunging in the water. I got up; but all was again quiet, except the cackle of fowls in our hen-coop, which hung over the sides of the vessel about three feet from the cabin door. I could

find no explanation of the circumstance, and, my men being all ashore, I turned in again and slept till morning. I then found my poultry loose about the canoe, and a large rent in the bottom of the hen-coop, which was about two feet from the surface of the water: a couple of fowls were missing. Senhor Antonio said the depredator was a Sucurujú (the Indian name for the Anaconda, or great water serpent — *Eunectes murinus*), which had for months past been haunting this part of the river, and had carried off many ducks and fowls from the ports of various houses. I was inclined to doubt the fact of a serpent striking at its prey from the water, and thought an alligator more likely to be the culprit, although we had not yet met with alligators in the river. Some days afterwards the young men belonging to the different sítios agreed together to go in search of the serpent. They began in a systematic manner, forming two parties each embarked in three or four canoes, and starting from points several miles apart, whence they gradually approximated, searching all the little inlets on both sides the river. The reptile was found at last sunning itself on a log at the mouth of a muddy rivulet, and despatched with harpoons. I saw it the day after it was killed: it was not a very large specimen, measuring only eighteen feet nine inches in length and sixteen inches in circumference at the widest part of the body. I measured skins of the Anaconda afterwards, twenty-one feet in length and two feet in girth. The reptile has a most hideous appearance, owing to its being very broad in the middle and tapering abruptly at both ends. It is very abundant in some parts of the country; nowhere more so than in the Lago Grande, near Santarém, where it is often seen coiled up in the corners of farmyards, and detested for its habit of carrying off poultry, young calves, or whatever animal it can get within reach of.

At Ega a large Anaconda was once near making a meal of a young lad about ten years of age belonging to one of my neighbours. The father and his son went one day in their montaria a few miles up the Teffé to gather wild fruit; landing on a sloping sandy shore, where the boy was left to mind the canoe whilst the man entered the forest. The beaches of the Teffé form groves of wild guava and myrtle trees, and during most months of the year are partly overflown by the river. Whilst the boy was playing in the water

under the shade of these trees a huge reptile of this species stealthily wound its coils around him, unperceived until it was too late to escape. His cries brought the father quickly to the rescue; who rushed forward, and seizing the Anaconda boldly by the head, tore his jaws asunder. There appears to be no doubt that this formidable serpent grows to an enormous bulk and lives to a great age, for I heard of specimens having been killed which measured forty-two feet in length, or double the size of the largest I had an opportunity of examining.

BUILD A CANOE

The situation was most favourable for collecting the natural products of the district. The forest was not crowded with underwood, and pathways led through it for many miles and in various directions. I could make no use here of our two men as hunters, so, to keep them employed whilst José and I worked daily in the woods, I set them to make a montaria under João Aracu's directions. The first day a suitable tree was found for the shell of the boat, of the kind called *Itauba amarello*, the yellow variety of the stone-wood. They felled it, and shaped out of the trunk a log nineteen feet in length: this they dragged from the forest, with the help of my host's men, over a road they had previously made with pieces of round wood to act as rollers. The distance was about half a mile, and the ropes used for drawing the heavy load were tough lianas cut from the surrounding trees. This part of the work occupied about a week: the log had then to be hollowed out, which was done with strong chisels through a slit made down the whole length. The heavy portion of the task being then completed, nothing remained but to widen the opening, fit two planks for the sides and the same number of semicircular boards for the ends, make the benches, and caulk the seams.

The expanding of the log thus hollowed out is a critical operation, and not always successful, many a good shell being spoilt by its splitting or expanding irregularly. It is first reared on tressels, with the slit downwards, over a large fire, which is kept up for seven or eight hours, the process requiring unremitting attention to avoid cracks and make the plank bend with the proper dip at the two ends. Wooden straddlers, made by cleaving

pieces of tough elastic wood and fixing them with wedges, are inserted into the opening, their compass being altered gradually as the work goes on, but in different degree according to the part of the boat operated upon. Our casca turned out a good one: it took a long time to cool, and was kept in shape whilst it did so by means of wooden cross-pieces. When the boat was finished it was launched with great merriment by the men, who hoisted coloured handkerchiefs for flags, and paddled it up and down the stream to try its capabilities. My people had suffered as much inconvenience from the want of a montaria as myself, so this was a day of rejoicing to all of us.

TIDES

A small creek traversed the forest behind João Aracu's house, and entered the river a few yards from our anchoring place. I used to cross it twice a day, on going and returning from my hunting ground. One day early in September, I noticed that the water was two or three inches higher in the afternoon than it had been in the morning. This phenomenon was repeated the next day, and in fact daily, until the creek became dry with the continued subsidence of the Cuparí, the time of rising shifting a little from day to day. I pointed out the circumstance to João Aracu, who had not noticed it before (it was only his second year of residence in the locality), but agreed with me that it must be the "mare." Yes, the tide! the throb of the great oceanic pulse felt in this remote corner, 530 miles distant from the place where it first strikes the body of fresh water at the mouth of the Amazons. I hesitated at first at this conclusion, but on reflecting that the tide was known to be perceptible at Obydos, more than 400 miles from the sea; that at high water in the dry season a large flood from the Amazons enters the mouth of the Tapajós, and that there is but a very small difference of level between that point and the Cuparí, a fact shown by the absence of current in the dry season; I could have no doubt that this conclusion was a correct one.

The fact of the tide being felt 530 miles up the Amazons, passing from the main stream to one of its affluents 380 miles from its mouth, and thence to a branch in the third degree, is a proof of the extreme flatness of the land which forms the lower part of the Amazonian valley.

Pumice Stones

The fishermen twice brought me small rounded pieces of very porous pumice-stone, which they had picked up floating on the surface of the main current of the river. They were to me objects of great curiosity as being messengers from the distant volcanoes of the Andes: Cotopaxi, Llanganete, or Sangay, which rear their peaks amongst the rivulets that feed some of the early tributaries of the Amazons, such as the Macas, the Pastaza, and the Napo. The stones must have already travelled a distance of 1,200 miles. I afterwards found them rather common: the Brazilians use them for cleaning rust from their guns, and firmly believe them to be solidified river foam. A friend once brought me, when I lived at Santarém, a large piece which had been found in the middle of the stream below Monte Alegre, about 900 miles further down the river: having reached this distance, pumice-stones would be pretty sure of being carried out to sea, and floated thence with the north-westerly Atlantic current to shores many thousand miles distant from the volcanoes which ejected them. They are sometimes found stranded on the banks in different parts of the river. Reflecting on this circumstance since I arrived in England, the probability of these porous fragments serving as vehicles for the transportation of seeds of plants, eggs of insects, spawn of fresh-water fish, and so forth, has suggested itself to me. Their rounded, water-worn appearance showed that they must have been rolled about for a long time in the shallow streams near the sources of the rivers at the feet of the volcanoes, before they leapt the waterfalls and embarked on the currents which lead direct for the Amazons. They may have been originally cast on the land and afterwards carried to the rivers by freshets; in which case the eggs and seeds of land insects and plants might be accidentally introduced and safely enclosed with particles of earth in their cavities. As the speed of the current in the rainy season has been observed to be from three to five miles an hour, they might travel an immense distance before the eggs or seeds were destroyed. I am ashamed to say that I neglected the opportunity, whilst on the spot, of ascertaining whether this was actually the case. The attention of Naturalists has only lately been turned to the important subject of occasional means of wide dissemination of species of animals

and plants. Unless such be shown to exist, it is impossible to solve some of the most difficult problems connected with the distribution of plants and animals. Some species, with most limited powers of locomotion, are found in opposite parts of the earth, without existing in the intermediate regions; unless it can be shown that these may have migrated or been accidentally transported from one point to the other, we shall have to come to the strange conclusion that the same species had been created in two separate districts.

Falling Banks

Canoemen on the Upper Amazons live in constant dread of the "terras cahidas," or landslips, which occasionally take place along the steep, earthy banks; especially when the waters are rising. Large vessels are sometimes overwhelmed by these avalanches of earth and trees. I should have thought the accounts of them exaggerated if I had not had an opportunity during this voyage of seeing one on a large scale. One morning I was awoke before sunrise by an unusual sound resembling the roar of artillery. I was lying alone on the top of the cabin; it was very dark, and all my companions were asleep, so I lay listening. The sounds came from a considerable distance, and the crash which had aroused me was succeeded by others much less formidable. The first explanation which occurred to me was that it was an earthquake; for, although the night was breathlessly calm, the broad river was much agitated and the vessel rolled heavily. Soon after, another loud explosion took place, apparently much nearer than the former one; then followed others. The thundering peal rolled backwards and forwards, now seeming close at hand, now far off; the sudden crashes being often succeeded by a pause or a long-continued dull rumbling. At the second explosion, Vicente, who lay snoring by the helm, awoke and told me it was a "terra cahida;" but I could scarcely believe him. The day dawned after the uproar had lasted about an hour, and we then saw the work of destruction going forward on the other side of the river, about three miles off. Large masses of forest, including trees of colossal size, probably 200 feet in height, were rocking to and fro, and falling headlong one after the other into the water. After each avalanche the wave which it caused returned on the

crumbly bank with tremendous force, and caused the fall of other masses by undermining them. The line of coast over which the landslip extended was a mile or two in length; the end of it, however, was hid from our view by an intervening island. It was a grand sight: each downfall created a cloud of spray; the concussion in one place causing other masses to give way a long distance from it, and thus the crashes continued, swaying to and fro, with little prospect of a termination.

MOUTH OF THE TEFFÉ

In the evening we arrived at a narrow opening, which would be taken by a stranger navigating the main channel for the outlet of some insignificant stream: it was the mouth of the Teffé, on whose banks Ega is situated, the termination of our voyage. After having struggled for 35 days with the muddy currents and insect pests of the Solimoens, it was unspeakably refreshing to find one's-self again in a dark-water river, smooth as a lake and free from Píum and Motúca. The rounded outline, small foliage, and sombre green of the woods, which seemed to rest on the glassy waters, made a pleasant contrast to the tumultuous piles of rank, glaring, light-green vegetation, and torn, timber-strewn banks to which we had been so long accustomed on the main river. The men rowed lazily until nightfall, when, having done a laborious day's work, they discontinued and went to sleep, intending to make for Ega in the morning. It was not thought worthwhile to secure the vessel to the trees or cast anchor, as there was no current. I sat up for two or three hours after my companions had gone to rest, enjoying the solemn calm of the night. Not a breath of air stirred; the sky was of a deep blue, and the stars seemed to stand forth in sharp relief; there was no sound of life in the woods, except the occasional melancholy note of some nocturnal bird. I reflected on my own wandering life: I had now reached the end of the third stage of my journey,and was now more than half way across the continent. It was necessary for me, on many accounts, to find a rich locality for Natural History explorations, and settle myself in it for some months or years. Would the neighbourhood of Ega turn out to be suitable, and should I, a solitary stranger on a strange errand, find a welcome amongst its people?

Insect Fauna
of
the Amazon Valley

A SELECTION OF PAGES FROM THE SECOND JOURNAL

A. Helicopis   3 species

Species

201   H. Gnidus

234 ♂

damp parts of forest & swampy places   more or less abundant
through't Amazons — Pará, Santarem, Ega

N 2   H. Cupido

442 ♂
443 ♀

Same places as No 1.   Ega —   also I believe same sp. at Pará

N 3   H. Cupido black ♂

444 ♂

Most likely same sp. as No 2 & probably the typical form of
have not now sufficient specimens to decide —   Ega —

G. Anteros

G. Anteros

(No 149)

(No 154)

(very fine spines beneath tarsi)

(No 156 & 157)

♂

♀
no perceptible
tarsal spines

♀
no perceptible
tarsal spines

(No 153)

♂

♀
♂

No 160

MESENE I.

Mesene Pharea

49 ♂
50 ♀ Scarlet species, Obd. birds

common throughout Amazons, settles underside leaves, wings flat expan old

Mes. leucophrys, Bat—

51 ♂ Aveyros
470 ♀ ? Villa nova —

scarlet, apical ⅓ fore wing also blk with a white spot middle nr bale margin

71 ♂ – Ega –
common at Aveyros

Mes. Hya

52 ♂
53 ♀

smaller than ♂ basal ⅓ hind border fore wings dolisk hind wing orange, rest of wing blk, fore wing
blk spot a width toward onto margin. ♀ yellow, basal ⅓ fore wing (except costa) yellow
antareen

— Bæotis Arope

54 ♂
55 ♀ ?

blk, basal third of hind wing carmine red or orange red ? — The ♀ I put on ground of above
found in same places as ♂

common throughout Amazons, same habits as last

6 ♂
7 ♀ ?

blk, a bell from mid of hind margin fore wing, across basal part of hind wing & the middle of
abdomen, orange yellow. ♀ I put with doubt on same grounds as No —
and throughout Amazons

80 ♂ Mesene Sophister Bat—

74 ♀ (Ega, rare)
sparing, rare

9 ♀

general throughout Amazons

The ♀ of this has the palpi long, visible from above, an exceptional case in this genus

G. Chrysogyra

(considered part of charis by Authors)

loons, gilded or silvery borders — neuration same as Nymph.

Charis, Symmachia &c.

ol

4948

not Cupari

no 2

466 ♂

ga, very rare

no 3

281 ♀

rest Cupari — (? ♀ of 466)

no 4

585 ♀

Ega. very rare

CHARIS

§ of **Charis**

Differs from Nymphidium more than most of the other genera of the
next Nymphidium group —

Body slender & short, wings elongate, nervation same

Head moderate, round, furnished with a cushion of dense, even soft scales; Palps
rather shorter than the face, not visible when viewed from above, rather close
rather parallel; 2nd jt clothed with elongate fine scales inclined upwards, & decumbent; term
in a mere conical point in both sexes. Antennae nearly as long as the body
club rather abrupt, elongate-oblong. Thorax ovate, equal in length to one half the
abdomen in both sexes. Fore wings elongate, triangular

Hind wings short rounded, anal angle not prominent & together with the outer border
scarcely passing apex of abdomen.

Legs rather thinly clothed with spread, fine pubescence.

(Nᵒ 173) ♂   Proboscis ½ the length
of abd
— else hispid =

(N°115)

EMESIS. I.

1 Emesis Mandana
5♂
5♀
(allau)   Flowers at skirts of woods Santarem, settling always with
wings expanded.

2 Emesis Fatimella Westw
17♂

was, skirts of woods, Santarem.

3   ♀ Fatimella
118♀ curious as differing from №2 only in shape of wings, the difference
being what usually obtains in other sex.—

Emesis Mandana
119♂

similar to №1, but more ruddy in colour

5 Emesis Lucinda Cr
0♂
1♀

Virgin forests — Pará, Cupari, flight quick, short, settles frequently
leaves, wings always extended — like the Emesis & true Nymphidium

— might also be included in genus Symmachia
        ♀ Chrysodyra

[illegible] species first Cupari

7— still more nearly approaching Symmachia than №6. in fact perhaps ♂ of №5

6♂ Cricosoma calligrapha Bates

very pretty little sp. — forest, Cupari

Nymphidiina

Symmachia

of next form to Emesis, perhaps only to be distinguished by
metallic lustre more or less of it scales

Nymphidina

G Anatole

Nymphidium, Boisduval

2nd Section — Species with great portion of their wings of White
colour, *or sometimes pale yellow*, with generally a marginal row of lunules to hind wings,
Fore wings & antennæ rather more elongate & slender

No 1

♂ 104
(yellow) Fore wings isabella-fulvous with black, transverse spots & lines
towards their base, hind wings apical half White, with marginal row
of black spots imperfect, basal p isabella-fulvous m

rather common throughout Amazons

No 2

♂ 105)
♀ 106)

Pará, Santarem.

No 3

♂ 914
♀ 915
(white)

thinned forest, Cupari

No 4

♂ 107
♀ 108 (yellow)

Very closely approximate to No 3, but not found in same localities, it is
larger & more ruddy in colour.

No 5

♂ 109

No 6

♂ 110
♀ 111

local — woods, Santarem

Nymphidiinæ

G. Tharops

4th Section   Species of bluish or metallic colours,

with blk spots & marks above   *Tharops* Doubleday

43 ♂
44 ♀
(♀♂)

handsome blue dull steel colour —   moist roots, Pará, Santarem, rare

145 ♂
yell.

Pará, unique

=
= 139 ♂

♀a, rare — & I think same sp. at Pará — prefers upper side of leaves &c

=
= 140 ♂

Villa Nova at flowers, believe same sp. at Pará

=
= 85 ♂

This sp. rather different in habits from the above, it does not frequent
the shade of the woods, nor is ever seen about foliage — it prefers
the muddy or sandy shores of rivers & sometimes is seen in open
places in forest: in all these places it settles on ground, at ordure
or moist chips of wood — delights also to settle on passing canoes
it is not common anywhere, I have seen it in the Cupari, at
Villa Nova & at Ega —

Nymphidium

g. Aricoris (?) males

Lemonias, Westwood

Aricoris

Body stout; Head rather large, broad, forehead & crown clothed
with dense, rather short, soft scales; eyes large, prominent.
Head larger in ♂ than in ♀; Palpi in ♂ elongate, thin tips,
conspicuous beyond the face when viewed from above, diverging
about their middle than approximating at their points; 2nd
jt. elongate clothed with short, dense, even scales; terminal scarcely
¼ th length of 2nd jt. much thinner, nearly naked & pointed; in ♀
Palpi much more elongate, the 2nd jt. passing the face by one
half their length, terminal jt. long naked, drooping. Antennæ
moderate, shorter than body, stout, club gradually thickening
from the base to the apex.

Thorax in ♂ short, oval, equal in length to 3/5 ths. of the abdomen.
Fore wings subtriangular, fore margin very slightly flexuous, apex
obtuse, outer margin generally slightly rounded, hind angle obtuse
& posterior margin straight. Costal nervure flexuous, stout, reaching
fore edge at abt. ½ length of the wing; discoidal cell terminating
about middle of the wing; first & second subcostal nervules arising
near together & both before the end of the cell, third arising about
half way between the end of the cell & the apex of the wing & with the
continuation of the subcostal nervure forming the final bifurcation
the fourth nervule thus being awanting. Upper discoidal
nervure arising from the subcostal nervure at the end of the cell, with a
very minute disco-cellular
disco-cellular nervule; middle & lower disco-cellulars of equal length
both rudimentary, the latter joining the median
nervure close after it final bifurcation & without forming an angle
& receive it. Hind wing in ♂ rather produced at anal angle
& rounded, outer margin nearly straight; in ♀ the anal angle not produced
but rounded regularly & broadly along outer margin.

Nymphidiinæ

6 Aricoris, 2 . females

§2 Section 2 –　　　　Apical border of hind wings in
both sexes festooned or produced into obtuse lobes at the termination of the submedian nervure & the 1st & 2nd median nervules –
Colours of ♀ resembling those of ♂ .　　　Habits same as Sect. 1
settling beneath leaves wing erect.

No 1
866 ♂
488 ♀

Splendid sp. at the Cupari & at Altar do Chão

Aricoris (?) females

Section 3 —   In Colours somewhat resembling some
of the Lemonias of Section 1st; but habits different, reposing
with wings extended.   The group is thus intermediate between
Anatole & Lemonias & would be equally well placed in the
one genus as in the other. The hind wings are more rounded
in the ♂ than in those of same sex of Sect. 1.

01 L. Irene Westw.

60♂

Ega, very rare

2. L. Siaka

618

magnificent species, doubtless new; Ega, rare

3

587♀

think the ♀ of 460 — found in same locality, in forest. Ega.

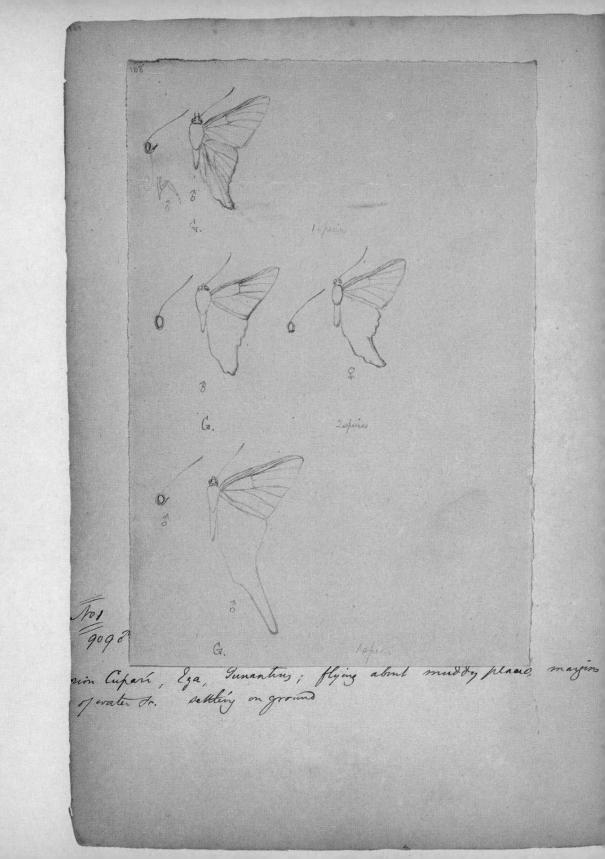

1 species

♂   ♀

G.   2 species

G.   1 species

No 1

9098

...rion Cupari, Ega, Tunantins; flying about muddy places, margins of water &c.   settling on ground

Syrmatia

one subcostal nervule before, one after the cell, antennae with
an abrupt, broad club. resembles Isapis nothing so ground

169

G. Isapis   2 species

G. Syrmatia   2 species

G   1 species

No 580   Isapis?

~ol
558 ♀

~ga, May 1857. flight like Leonia

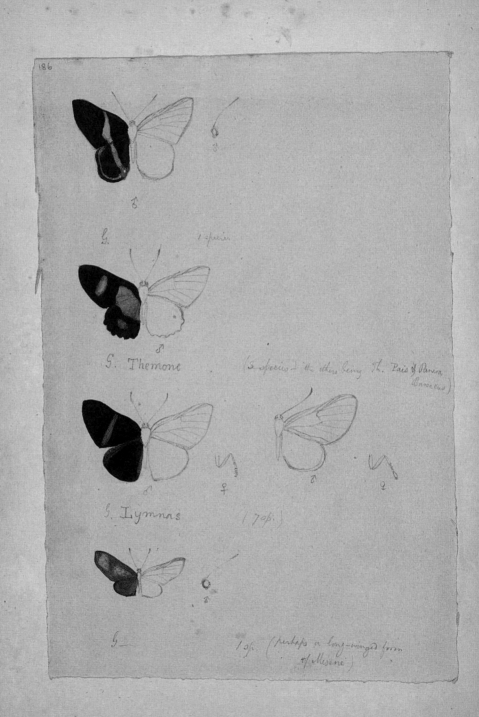

186

♂

G.   1 species

♂

S. Themone   (2 species) the other being M. Pais & Panem (Boisduval)

G. Lymnas   (7 sp.)

♂   ♀   ♂   ♀

G.   1 sp. (perhaps a long-winged form of Mesene)

Erycinidæ

Themone trestri.

Head moderate, orbicular; forehead clothed with short, soft scales
scarcely prominent; palpi shorter than the face, closely clothed with short
compact, even scales, terminal its short pointed

Antennæ short, club oblong, large, solid, abruptly thickening from
the stem.

Shape of wings similar to the Nymphidia. Nemation
on the plan of the Erycina veræ, almost identical
with Panara, Isapis &c. & differing from all the Nymphidiinæ in having the first subcostal branch arising
before & 2nd after the end of cell. The shape of
cell & straightness of subcostal nervure very similar
to the Nymphidiinæ. No upper discocellular in
fore wings, their origin of the middle discocell. being in conjunction with that of upper discoidal. The
lower discocellular is tubular for a portion of its distance from the median. The
rest of the discocellular rudimentary. The fore leg ♂ very feeble, uniarticula-
tarsus, clad with longish hairs. other legs moderate, femora banded beneath, tarsi
with very small spines &c. Palpi short and slender as fig. Proboscis
not & very slender. antennæ strongly clavate as fig. (Ega 10 April 1857)

No. 456

Not sure this sp. is the Ioapis Agyotus. Fore wings elongate - hind wings short.
1st subcostal nervule emitted before, 2nd after, the end of the cell, the latter
at a long distance after the cell. No upper disco-cellular nervule, the 1st
discoidal nervure continuing nearly straight with the basal part of the
subcostal. Middle & lower disco-cellulars continuous & apparently tubu-
but faintly marked & not corneous. Foreleys ♀ with basal jt. of tarsi as lo
as rest united, 2nd jt. very broad - Claw jt. tapering, claws not visible wi
simple lens. 2nd to 4th jts en. with a single spine beneath. Palpi ve
short wholly inferior.

Not ~~Agyotus Euo~~ Chamelimnas Iaeris
455 ♂
456 ♀
Thinned forest, Ega, flying very low near ground & feebly, settling undersid
leaves wings expanded as Lymnas. —

Erycinina ?
G. Othomiada

Erycinina.   Genus allied to Lymnas, Themore
                                                      12

The Naturalist on the River Amazons

EGA AND THE UPPER AMAZONS

I MADE EGA my headquarters during the whole of the time I remained on the Upper Amazons (four years and a half). My excursions into the neighbouring region extended sometimes as far as 300 and 400 miles from the place... In the intervals between them I led a quiet, uneventful life in the settlement; following my pursuit in the same peaceful, regular way as a Naturalist might do in a European village. For many weeks in succession my journal records little more than the notes made on my daily captures. I had a dry and spacious cottage, the principal room of which was made a workshop and study; here a large table was placed, and my little library of reference arranged on shelves in rough wooden boxes. Cages for drying specimens were suspended from the rafters by cords well anointed, to prevent ants from descending, with a bitter vegetable oil: rats and mice were kept from them by inverted cuyas, placed half-way down the cords. I always kept on hand a large portion of my private collection, which contained a pair of each species and variety, for the sake of comparing the old with the new acquisitions. My cottage was whitewashed inside and out about once a year by the proprietor, a native trader; the floor was of earth; the ventilation was perfect, for the outside air, and sometimes the rain as well, entered freely through gaps at the top of the walls under the eaves and through wide crevices in the doorways. Rude as the dwelling was, I look back with pleasure on the many happy months I spent in it. I rose generally with the sun, when the grassy streets were wet with dew, and walked down to the

river to bathe: five or six hours of every morning were spent collecting in the forest, whose borders lay only five minutes' walk from my house: the hot hours of the afternoon and the rainy days, were occupied in preparing and ticketing the specimens, making notes, dissecting, and drawing. I frequently had short rambles by water in a small montaria, with an Indian lad to paddle. The neighbourhood yielded me, up to the last day of my residence, an uninterrupted succession of new and curious forms in the different classes of the animal kingdom, but especially insects.

I lived, as may already have been seen, on the best of terms with the inhabitants of Ega. Refined society, of course, there was none; but the score or so of decent, quiet families which constituted the upper class of the place were very sociable; their manners offered a curious mixture of naive rusticity and formal politeness; the great desire to be thought civilised leads the most ignorant of these people (and they are all very ignorant, although of quick intelligence) to be civil and kind to strangers from Europe. I was never troubled with that impertinent curiosity on the part of the people in these interior places which some travellers complain of in other countries. The Indians and lower half-castes — at least such of them who gave any thought to the subject — seemed to think it natural that strangers should collect and send abroad the beautiful birds and insects of their country. The butterflies they universally concluded to be wanted as patterns for bright-coloured calico-prints. As to the better sort of people, I had no difficulty in making them understand that each European capital had a public museum, in which were sought to be stored specimens of all natural productions in the mineral, animal, and vegetable kingdoms. They could not comprehend how a man could study science for its own sake; but I told them I was collecting for the "Museo de Londres," and was paid for it; that was very intelligible. One day, soon after my arrival, when I was explaining these things to a listening circle seated on benches in the grassy street, one of the audience, a considerable tradesman, a Mameluco native of Ega, got suddenly quite enthusiastic, and exclaimed "How rich are these great nations of Europe! We half-civilised creatures know nothing. Let us treat this stranger well, that he may stay amongst us and teach our children."

Alligators

Alligators were rather troublesome in the dry season. During these months there was almost always one or two lying in wait near the bathing-place for anything that might turn up at the edge of the water; dog, sheep, pig, child, or drunken Indian. When this visitor was about, every one took extra care whilst bathing. I used to imitate the natives in not advancing far from the bank and in keeping my eye fixed on that of the monster, which stares with a disgusting leer along the surface of the water; the body being submerged to the level of the eyes, and the top of the head, with part of the dorsal crest, the only portions visible. When a little motion was perceived in the water behind the reptile's tail, bathers were obliged to beat a quick retreat. I was never threatened myself, but I often saw the crowds of women and children scared whilst bathing by the beast making a movement towards them; a general scamper to the shore and peals of laughter were always the result in these cases. The men can always destroy these alligators when they like to take the trouble to set out with montarias and harpoons for the purpose, but they never do it unless one of the monsters, bolder than usual, puts someone's life in danger. This arouses them, and they then track the enemy with the greatest pertinacity; when half killed they drag it ashore and despatch it amid loud execrations. Another, however, is sure to appear some days or weeks afterwards, and take the vacant place on the station. Besides alligators, the only animals to be feared are the poisonous serpents. These are certainly common enough in the forest, but no accident happened during the whole time of my residence.

Ypadú

The half-caste and Indian women, after middle age, are nearly all addicted to the use of Ypadú, the powdered leaves of a plant (*Erythroxylon coca*) which is well known as a product of the eastern parts of Peru, and is to the natives of these regions what opium is to the Turks and betel to the Malays. Persons who indulge in Ypadú at Ega are held in such abhorrence, that they keep the matter as secret as possible; so it is said, and no doubt with truth, that the slender result of the women's daily visits to their roças,

is owing to their excessive use of this drug. They plant their little plots of the tree in retired nooks in the forest, and keep their stores of the powder in hiding-places near the huts which are built on each plantation. Taken in moderation, Ypadú has a stimulating and not injurious effect, but in excess it is very weakening, destroying the appetite, and producing in time great nervous exhaustion. I once had an opportunity of seeing it made at the house of a Maraud Indian on the banks of the Jutahi. The leaves were dried on a mandioca oven, and afterwards pounded in a very long and narrow wooden mortar. When about half pulverised, a number of the large leaves of the *Cecropia palmata* (candelabrum tree) were burnt on the floor, and the ashes dirtily gathered up and mixed with the powder. The Ypadú-eaters say that this prevents the ill-effects which would arise from the use of the pure leaf, but I should think the mixture of so much indigestible filth would be more likely to have the opposite result.

Turtle

We lived at Ega, during most part of the year, on turtle. The great fresh-water turtle of the Amazons grows on the upper river to an immense size, a full-grown one measuring nearly three feet in length by two in breadth, and is a load for the strongest Indian. Every house has a little pond, called a curral (pen), in the back-yard to hold a stock of the animals through the season of dearth — the wet months; those who have a number of Indians in their employ sending them out for a month when the waters are low, to collect a stock, and those who have not, purchasing their supply; with some difficulty, however, as they are rarely offered for sale. The price of turtles, like that of all other articles of food, has risen greatly with the introduction of steam-vessels. When I arrived in 1850 a middle-sized one could be bought pretty readily for ninepence, but when I left in 1859, they were with difficulty obtained at eight and nine shillings each. The abundance of turtles, or rather the facility with which they can be found and caught, varies with the amount of annual subsidence of the waters. When the river sinks less than the average, they are scarce; but when more, they can be caught in plenty, the bays and shallow lagoons in the forest having then only a small depth

of water. The flesh is very tender, palatable, and wholesome; but it is very cloying: everyone ends, sooner or later, by becoming thoroughly surfeited. I became so sick of turtle in the course of two years that I could not bear the smell of it, although at the same time nothing else was to be had, and I was suffering actual hunger. The native women cook it in various ways. The entrails are chopped up and made into a delicious soup called sara-patel, which is generally boiled in the concave upper shell of the animal used as a kettle. The tender flesh of the breast is partially minced with farinha, and the breast shell then roasted over the fire, making a very pleasant dish. Steaks cut from the breast and cooked with the fat form another palatable dish. Large sausages are made of the thick-coated stomach, which is filled with minced meat and boiled. The quarters cooked in a kettle of Tucupi sauce form another variety of food. When surfeited with turtle in all other shapes, pieces of the lean part roasted on a spit and moistened only with vinegar make an agreeable change. The smaller kind of turtle, the Tracajá, which makes its appearance in the main river, and lays its eggs a month earlier than the large species, is of less utility to the inhabitants although its flesh is superior, on account of the difficulty of keeping it alive; it survives captivity but a very few days, although placed in the same ponds in which the large turtle keeps well for two or three years.

Excursions around Ega

Every tree was tenanted by Cicadas, the reedy notes of which produced that loud, jarring, insect music which is the general accompaniment of a woodland ramble in a hot climate. One species was very handsome, having wings adorned with patches of bright green and scarlet. It was very common; sometimes three or four tenanting a single tree, clinging as usual to the branches. On approaching a tree thus peopled, a number of little jets of a clear liquid would be seen squirted from aloft. I have often received the well-directed discharge full on my face; but the liquid is harmless, having a sweetish taste, and is ejected by the insect from the anus, probably in self-defence, or from fear. The number and variety of gaily-tinted butterflies, sporting about in this grove on sunny days, were so great that the bright

moving flakes of colour gave quite a character to the physiognomy of the place. It was impossible to walk far without disturbing flocks of them from the damp sand at the edge of the water, where they congregated to imbibe the moisture. They were of almost all colours, sizes, and shapes: I noticed here altogether eighty species, belonging to twenty-two different genera. It is a singular fact that, with very few exceptions, all the individuals of these various species thus sporting in sunny places were of the male sex; their partners, which are much more soberly dressed and immensely less numerous than the males, being confined to the shades of the woods. Every afternoon, as the sun was getting low, I used to notice these gaudy sunshine-loving swains trooping off to the forest, where I suppose they would find their sweethearts and wives. The most abundant, next to the very common sulphur-yellow and orange-coloured kinds (*Callidryas*, seven species), were about a dozen species of *Cybdelis*, which are of large size, and are conspicuous from their liveries of glossy dark-blue and purple. A superbly-adorned creature, the *Callithea markii*, having wings of a thick texture, coloured sapphire-blue and orange, was only an occasional visitor. On certain days, when the weather was very calm, two small gilded-green species (*Symmachia trochilus* and *S. colubris*) literally swarmed on the sands, their glittering wings lying wide open on the flat surface. The beach terminates, eight miles beyond Ega, at the mouth of a rivulet; the character of the coast then changes, the river banks being masked by a line of low islets amid a labyrinth of channels.

Blowpipe

This instrument is used by all the Indian tribes on the Upper Amazons. It is generally nine or ten feet long, and is made of two separate lengths of wood, each scooped out so as to form one half of the tube. To do this with the necessary accuracy requires an enormous amount of patient labour, and considerable mechanical ability, the tools used being simply the incisor teeth of the Páca and Cutía. The two half tubes, when finished, are secured together by a very close and tight spirally wound strapping, consisting of long flat strips of Jacitára, or the wood of the climbing palm-tree; and the

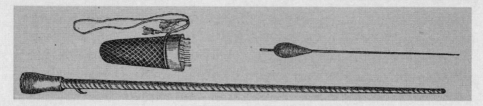

Blowpipe, quiver and arrow

whole is smeared afterwards with black wax, the production of a *Melipona* bee. The pipe tapers towards the muzzle, and a cup-shaped mouthpiece, made of wood, is fitted in the broad end. A full-sized Zarabatana is heavy, and can only be used by an adult Indian who has had great practice. The young lads learn to shoot with smaller and lighter tubes. When Mr. Wallace and I had lessons at Barra in the use of the blowpipe, of Julio, a Jurí Indian, then in the employ of Mr. Hauxwell, an English bird collector, we found it very difficult to hold steadily the long tubes. The arrows are made from the hard rind of the leaf-stalks of certain palms, thin strips being cut, and rendered as sharp as needles by scraping the ends with a knife or the tooth of an animal. They are winged with a little oval mass of samaúma silk (from the seed-vessels of the silk-cotton tree, *Eriodendron samauma*), cotton being too heavy. The ball of samaúma should fit to a nicety the bore of the blowpipe; when it does so, the arrow can be propelled with such force by the breath that it makes a noise almost as loud as a pop-gun on flying from the muzzle. My little companion was armed with a quiver full of these little missiles, a small number of which, sufficient for the day's sport, were tipped with the fatal Urari poison. The quiver was an ornamental affair, the broad rim being made of highly-polished wood of a rich cherry-red colour (the Moira-piránga, or red-wood of the Japurá). The body was formed of neatly-plaited strips of Maranta stalks, and the belt by which it was suspended from the shoulder was decorated with cotton fringes and tassels.

We walked about two miles along a well-trodden pathway, through high caäpoeira (second-growth forest). A large proportion of the trees were *Melastomas*, which bore a hairy yellow fruit, nearly as large and as well flavoured as our gooseberry. The season, however, was nearly over for them.

The road was bordered every inch of the way by a thick bed of elegant *Lycopodiums.* An artificial arrangement of trees and bushes could scarcely have been made to wear so finished an appearance as this naturally decorated avenue. The path at length terminated at a plantation of mandioca, the largest I had yet seen since I left the neighbourhood of Pará. There were probably ten acres of cleared land, and part of the ground was planted with Indian corn, water-melons, and sugar-cane. Beyond this field there was only a faint hunter's track, leading towards the untrodden interior. My companion told me he had never heard of there being any inhabitants in that direction (the south). We crossed the forest from this place to another smaller clearing, and then walked, on our road home, through about two miles of caäpoeira of various ages, the sites of old plantations. The only fruits of our ramble were a few rare insects and a Japú (*Cassicus cristatus*), a handsome bird with chestnut and saffron-coloured plumage, which wanders through the tree-tops in large flocks. My little companion brought this down from a height which I calculated at thirty yards. The blowpipe, however, in the hands of an expert adult Indian, can be made to propel arrows so as to kill at a distance of fifty and sixty yards. The aim is most certain when the tube is held vertically, or nearly so. It is a far more useful weapon in the forest than a gun, for the report of a firearm alarms the whole flock of birds or monkeys feeding on a tree, whilst the silent poisoned dart brings the animals down one by one until the sportsman has a heap of slain by his side. None but the stealthy Indian can use it effectively. The poison, which must be fresh to kill speedily, is obtained only of the Indians who live beyond the cataracts of the rivers flowing from the north, especially the Rio Negro and the Japurá. Its principal ingredient is the wood of the *Strychnos toxifera*, a tree which does not grow in the humid forests of the river plains.

Morning on the Praia

On rising I went to join my friends. Few recollections of my Amazonian rambles are more vivid and agreeable than that of my walk over the white sea of sand on this cool morning. The sky was cloudless; the just-risen sun was hidden behind the dark mass of woods on Shimuni, but the long line

of forest to the west, on Baria, with its plumy decorations of palms, was lighted up with his yellow, horizontal rays. A faint chorus of singing birds reached the ears from across the water, and flocks of gulls and plovers were crying plaintively over the swelling banks of the praia, where their eggs lay in nests made in little hollows of the sand. Tracks of stray turtles were visible on the smooth white surface of the praia. The animals which thus wander from the main body are lawful prizes of the sentinels; they had caught in this way two before sun-rise, one of which we had for dinner. In my walk I disturbed several pairs of the chocolate and drab-coloured wild goose (*Anser jubatus*) which set off to run along the edge of the water. The enjoyment one feels in rambling over these free, open spaces, is no doubt enhanced by the novelty of the scene, the change being very great from the monotonous landscape of forest which everywhere else presents itself.

On arriving at the edge of the forest I mounted the sentinel's stage, just in time to see the turtles retreating to the water on the opposite side of the sand-bank, after having laid their eggs. The sight was well worth the trouble of ascending the shaky ladder. They were about a mile off, but the surface of the sands was blackened with the multitudes which were waddling towards the river; the margin of the praia was rather steep, and they all seemed to tumble head first down the declivity into the water.

Bivouac on the Sand Bank

We dined on the banks of the river, a little before sunset. The mosquitoes then began to be troublesome, and finding it would be impossible to sleep here, we all embarked and crossed the river to a sand-bank, about three miles distant, where we passed the night. Cardozo and I slept in our hammocks slung between upright poles, the rest stretching themselves on the sand round a large fire. We lay awake conversing until past midnight. It was a real pleasure to listen to the stories told by one of the older men, they were given with so much spirit. The tales always related to struggles with some intractable animal — jaguar, manatee, or alligator. Many interjections and expressive gestures were used, and at the end came a sudden "Pa! terra!" when the animal was vanquished by a shot or a blow. Many mysterious

tales were recounted about the Bouto, as the large Dolphin of the Amazons is called. One of them was to the effect that a Bouto once had the habit of assuming the shape of a beautiful woman, with hair hanging loose to her heels, and walking ashore at night in the streets of Ega, to entice the young men down to the water. If anyone was so much smitten as to follow her to the water-side, she grasped her victim round the waist and plunged beneath the waves with a triumphant cry. No animal in the Amazons region is the subject of so many fables as the Bouto; but it is probable these did not originate with the Indians but with the Portuguese colonists.

Sports on the Praia

It was not all work at Catuá; indeed there was rather more play than work going on. The people make a kind of holiday of these occasions. Every fine night parties of the younger people assembled on the sands, and dancing and games were carried on for hours together. But the requisite liveliness for these sports was never got up without a good deal of preliminary rum-drinking. The girls were so coy that the young men could not get sufficient partners for the dances, without first subscribing for a few flagons of the needful cashaça. The coldness of the shy Indian and Mameluco maidens never failed to give way after a little of this strong drink, but it was astonishing what an immense deal they could take of it in the course of an evening. Coyness is not always a sign of innocence in these people, for most of the half-caste women on the Upper Amazons lead a little career of looseness before they marry and settle down for life; and it is rather remarkable that the men do not seem to object much to their brides having had a child or two by various fathers before marriage. The women do not lose reputation unless they become utterly depraved, but in that case they are condemned pretty strongly by public opinion. Depravity is, however, rare, for all require more or less to be wooed before they are won. I did not see (although I mixed pretty freely with the young people) any breach of propriety on the praias. The merrymakings were carried on near the ranchos, where the more staid citizens of Ega, husbands with their wives and young daughters, all smoking gravely out of long pipes, sat in their hammocks and enjoyed

the fun. Towards midnight we often heard, in the intervals between jokes and laughter, the hoarse roar of jaguars prowling about the jungle in the middle of the praia. There were several guitar players amongst the young men, and one most persevering fiddler, so there was no lack of music.

The favourite sport was Pira-purasséya, or fishdance, an original game of the Indians, though probably a little modified. The young men and women, mingling together, formed a ring, leaving one of them in the middle, who represented the fish. They then all marched round, Indian file, the musicians mixed up with the rest, singing a monotonous but rather pretty chorus, the words of which were invented by one of the party who acted as leader. This finished, all joined hands, and questions were put to the one in the middle, asking what kind of fish he or she might be. To these the individual has to reply. The end of it all is that he makes a rush at the ring, and if he succeeds in escaping, the person who allowed him to do so has to take his place; the march and chorus then recommence, and so the game goes on hour after hour. Tupi was the language mostly used, but sometimes Portuguese was sung and spoken. The details of the dance were often varied. Instead of the names of fishes being called over by the person in the middle, the name of some animal, flower, or other object was given to every fresh occupier of the place. There was then good scope for wit in the invention of nicknames, and peals of laughter would often salute some particularly good hit. Thus a very lanky young man was called the Magoary, or the grey stork; a moist grey-eyed man with a profile comically suggestive of a fish was christened Jaraki (a kind of fish), which was considered quite a witty sally; a little Mameluco girl, with light-coloured eyes and brown hair, got the gallant name of Rosa branca, or the white rose; a young fellow who had recently singed his eyebrows by the explosion of fireworks was dubbed Pedro queimado (burnt Peter); in short everyone got a nickname, and each time the cognomen was introduced into the chorus as the circle marched round.

UMBRELLA BIRD

Umbrella Bird (*Cephalopterus ornatus*), a species which resembles in size, colour, and appearance our common crow, but is decorated with a crest of

Umbrella bird

long, curved, hairy feathers having long bare quills, which, when raised, spread themselves out in the form of a fringed sun-shade over the head. A strange ornament, like a pelerine, is also suspended from the neck, formed by a thick pad of glossy steel-blue feathers, which grow on a long fleshy lobe or excrescence. This lobe is connected (as I found on skinning specimens) with an unusual development of the trachea and vocal organs, to which the bird doubtless owes its singularly deep, loud, and long-sustained fluty note. The Indian name of this strange creature is Uirá-mimbéu, or fife-bird, in

allusion to the tone of its voice. We had the good luck, after remaining quiet a short time, to hear its performance. It drew itself up on its perch, spread widely the umbrella-formed crest, dilated and waved its glossy breast-lappet, and then, in giving vent to its loud piping note, bowed its head slowly forwards. We obtained a pair, male and female: the female has only the rudiments of the crest and lappet, and is duller-coloured altogether than the male. The range of this bird appears to be quite confined to the plains of the Upper Amazons (especially the Ygapó forests), not having been found to the east of the Rio Negro.

Ygapó Forest

After walking about half a mile we came upon a dry water-course, where we observed, first, the old footmarks of a tapir, and, soon after, on the margins of a curious circular hole full of muddy water, the fresh tracks of a Jaguar. This latter discovery was hardly made, when a rush was heard amidst the bushes on the top of a sloping bank on the opposite side of the dried creek. We bounded forward; it was, however, too late, for the animal had sped in a few moments far out of our reach. It was clear we had disturbed, on our approach, the Jaguar, whilst quenching his thirst at the water-hole. A few steps further on we saw the mangled remains of an alligator (the Jacaretinga). The head, fore-quarters, and bony shell were the only parts which remained; but the meat was quite fresh, and there were many footmarks of the Jaguar around the carcase; so that there was no doubt this had formed the solid part of the animal's breakfast. My companions now began to search for the alligator's nest, the presence of the reptile so far from the river being accountable for on no other ground than its maternal solicitude for its eggs. We found, in fact, the nest at the distance of a few yards from the place. It was a conical pile of dead leaves, in the middle of which twenty eggs were buried. These were of elliptical shape, considerably larger than those of a duck, and having a hard shell of the texture of porcelain, but very rough on the outside. They make a loud sound when rubbed together, and it is said that it is easy to find a mother alligator in the Ygapó forests, by rubbing together two eggs in this way, she being never far off, and attracted by the sounds.

I put half-a-dozen of the alligator's eggs in my game-bag for specimens, and we then continued on our way. Lino, who was now first, presently made a start backwards, calling out "Jararaca!" This is the name of a poisonous snake (genus *Craspedocephalus*), which is far more dreaded by the natives than Jaguar or Alligator. The individual seen by Lino lay coiled up at the foot of a tree, and was scarcely distinguishable, on account of the colours of its body being assimilated to those of the fallen leaves. Its hideous, flat triangular head, connected with the body by a thin neck, was reared and turned towards us: Frazao killed it with a charge of shot, shattering it completely, and destroying, to my regret, its value as a specimen. In conversing on the subject of Jararacas as we walked onwards, every one of the party was ready to swear that this snake attacks man without provocation, leaping towards him from a considerable distance when he approaches. I met, in the course of my daily rambles in the woods, many Jararacas, and once or twice very narrowly escaped treading on them, but never saw them attempt to spring. On some subjects the testimony of the natives of a wild country is utterly worthless. The bite of the Jararacas is generally fatal. I knew of four or five instances of death from it, and only of one clear case of recovery after being bitten; but in that case the person was lamed for life.

We walked over moderately elevated and dry ground for about a mile, and then descended (three or four feet only) to the dry bed of another creek. This was pierced in the same way as the former water-course, with round holes full of muddy water. They occurred at intervals of a few yards, and had the appearance of having been made by the hand of man. The smallest were about two feet, the largest seven or eight feet in diameter. As we approached the most considerable of the larger ones, I was startled at seeing a number of large serpent-like heads bobbing above the surface. They proved to be those of electric eels, and it now occurred to me that these round holes were made by these animals working constantly round and round in the moist muddy soil. Their depth (some of them were at least eight feet deep) was doubtless due also to the movements of the eels in the soft soil, and accounted for their not drying up, in the fine season, with the rest of the creek. Thus, whilst alligators and turtles in this great inundated

forest region retire to the larger pools during the dry season, the electric
eels make for themselves little ponds in which to pass the season of drought.

My companions now cut each a stout pole, and proceeded to eject the
eels in order to get at the other fishes, with which they had discovered the
ponds to abound. I amused them all very much by showing how the electric
shock from the eels could pass from one person to another. We joined hands
in a line whilst I touched the biggest and freshest of the animals on the
head with the point of my hunting-knife. We found that this experiment
did not succeed more than three times with the same eel when out of the
water: for, the fourth time, the shock was scarcely perceptible.

BATS

The only other mammals that I shall mention are the bats, which exist
in very considerable numbers and variety in the forest, as well as in the
buildings of the villages. Many small and curious species living in the
woods, conceal themselves by day under the broad leaf-blades of Heliconiae
and other plants which grow in shady places; others cling to the trunks of
trees. Whilst walking through the forest in the daytime, especially along
gloomy ravines, one is almost sure to startle bats from their sleeping-places;
and at night they are often seen in great numbers flitting about the trees
on the shady margins of narrow channels. I captured altogether, without
giving especial attention to bats, sixteen different species at Ega.

The Vampire Bat. — The little grey bloodsucking *Phyllostoma*.... was
not uncommon at Ega, where everyone believes it to visit sleepers and bleed
them in the night. But the vampire was here by far the most abundant of
the family of leaf-nosed bats. It is the largest of all the South American
species, measuring twenty-eight inches in expanse of wing. Nothing, in
animal physiognomy can be more hideous than the countenance of this
creature when viewed from the front; the large, leathery ears standing out
from the sides and top of the head, the erect spear-shaped appendage on the
tip of the nose, the grin and the glistening black eye all combining to make
up a figure that reminds one of some mocking imp of fable. No wonder
that imaginative people have inferred diabolical instincts on the part of

so ugly an animal. The vampire, however, is the most harmless of all bats, and its inoffensive character is well known to residents on the banks of the Amazons. I found two distinct species of it, one having the fur of a blackish colour, the other of a ruddy hue, and ascertained that both feed chiefly on fruits. The church at Ega was the head-quarters of both kinds; I used to see them, as I sat at my door during the short evening twilights, trooping forth by scores from a large open window at the back of the altar, twittering cheerfully as they sped off to the borders of the forest. They sometimes enter houses; the first time I saw one in my chamber, wheeling heavily round and round, I mistook it for a pigeon, thinking that a tame one had escaped from the premises of one of my neighbours. I opened the stomachs of several of these bats, and found them to contain a mass of pulp and seeds of fruits, mingled with a few remains of insects. The natives say they devour ripe cajus and guavas on trees in the gardens, but on comparing the seeds taken from their stomachs with those of all cultivated trees at Ega, I found they were unlike any of them; it is therefore probable that they generally resort to the forest to feed, coming to the village in the morning to sleep, because they find it more secure from animals of prey than their natural abodes in the woods.

INSECTS

Upwards of 7,000 species of insects were found in the neighbourhood of Ega. I must confine myself, in this place, to a few remarks on the order Lepidoptera, and on the ants, several kinds of which, found chiefly on the Upper Amazons, exhibit the most extraordinary instincts.

I have mentioned, in a former chapter, the general sultry condition of the atmosphere on the Upper Amazons, where the sea-breezes which blow from Pará to the mouth of the Rio Negro (1,000 miles up stream) are unknown. This simple difference of meteorological conditions would hardly be thought to determine what genera of butterflies should inhabit each region, yet it does so in a very decisive manner. The Upper Amazons, from Ega upwards, and the eastern slopes of the Andes, whence so large a number of the most richly-coloured species of this tribe have been received

in Europe, owe the most ornamental part of their insect population to the absence of strong and regular winds. Nineteen of the most handsome genera of Ega, containing altogether about 100 species, are either entirely absent or very sparingly represented on the Lower Amazons within reach of the trade winds. The range of these nineteen genera is affected by a curiously complicated set of circumstances. In all the species of which they are composed, the males are more than 100 to one more numerous than the females, and being very richly coloured, whilst the females are of dull hues, they spend their lives in sporting about in the sunlight, imbibing the moisture which constitutes their food, from the mud on the shores of streams, their spouses remaining hid in the shades of the forest. The very existence of these species depends on the facilities which their males have for indulgence in the pleasures of this sunshiny life. The greatest obstacle to this is the prevalence of strong winds, which not only dries rapidly all moisture in open places, but prevents the richly-attired dandies from flying daily to their feeding-places. I noticed this particularly whilst residing at Santarém, where the moist margins of water, localities which on the Upper Amazons swarm with these insects, were nearly destitute of them; and at Villa Nova (where a small number exists) I have watched them buffeting with the strong winds at the commencement of the dry season, and, as the dryness increased, disappearing from the locality. On ascending the Tapajós to the calm and sultry banks of the Cuparí, a great number of these insects reappeared, most of them being the same as those found on the Upper Amazons, thus showing clearly that their existence in the district depended on the absence of winds.

Before proceeding to describe the ants, a few remarks may be made on the singular cases and cocoons woven by the caterpillars of certain moths found at Ega. The first that may be mentioned, is one of the most beautiful examples of insect workmanship I ever saw. It is a cocoon, about the size of a sparrow's egg, woven by a caterpillar in broad meshes of either buff or rose-coloured silk, and is frequently seen in the narrow alleys of the forest, suspended from the extreme tip of an outstanding leaf by a strong silken thread five or six inches in length. It forms a very conspicuous object,

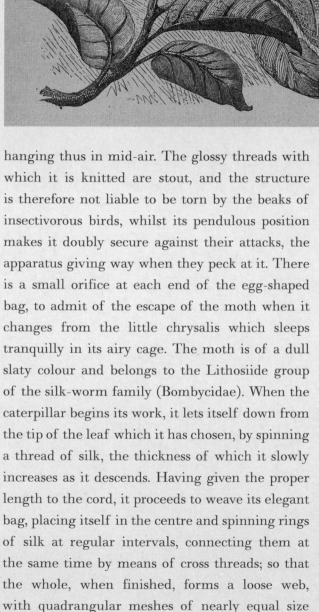

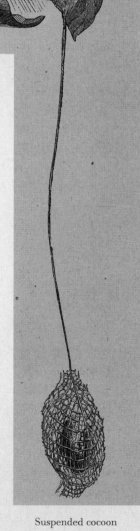

hanging thus in mid-air. The glossy threads with which it is knitted are stout, and the structure is therefore not liable to be torn by the beaks of insectivorous birds, whilst its pendulous position makes it doubly secure against their attacks, the apparatus giving way when they peck at it. There is a small orifice at each end of the egg-shaped bag, to admit of the escape of the moth when it changes from the little chrysalis which sleeps tranquilly in its airy cage. The moth is of a dull slaty colour and belongs to the Lithosiide group of the silk-worm family (Bombycidae). When the caterpillar begins its work, it lets itself down from the tip of the leaf which it has chosen, by spinning a thread of silk, the thickness of which it slowly increases as it descends. Having given the proper length to the cord, it proceeds to weave its elegant bag, placing itself in the centre and spinning rings of silk at regular intervals, connecting them at the same time by means of cross threads; so that the whole, when finished, forms a loose web, with quadrangular meshes of nearly equal size throughout. The task occupies about four days: when finished, the enclosed caterpillar becomes

Suspended cocoon
of moth

sluggish, its skin shrivels and cracks, and there then remains a motionless chrysalis of narrow shape, leaning against the sides of its silken cage.

Many other kinds are found at Ega belonging to the same cocoon-weaving family, some of which differ from the rest in their caterpillars possessing the art of fabricating cases with fragments of wood or leaves, in which they live secure from all enemies whilst they are feeding and growing. I saw many species of these; some of them knitted together, with fine silken threads, small bits of stick, and so made tubes similar to those of caddice-worms; others (*Saccophora*) chose leaves for the same purpose, forming with them an elongated bag open at both ends, and having the inside lined with a thick web. The tubes of full-grown caterpillars of *Saccophora* are two inches in length, and it is at this stage of growth, that I have generally seen them. They feed on the leaves of *Melastomse*, and as, in crawling, the weight of so large a dwelling would be greater than the contained caterpillar could sustain, the insect attaches the case by one or more threads to the leaves or twigs near which it is feeding.

Foraging Ants

Many confused statements have been published in books of travel, and copied in Natural History works, regarding these ants, which appear to have been confounded with the Saüba, a sketch of whose habits has been given in the first chapter of this work. The Saüba is a vegetable feeder, and does not attack other animals; the accounts that have been published regarding carnivorous ants which hunt in vast armies, exciting terror wherever they go, apply only to the *Ecitons*, or foraging ants, a totally different group of this tribe of insects. The *Ecitons* are called Tauoca by the Indians, who are always on the look-out for their armies when they traverse the forest, so as to avoid being attacked. I met with ten distinct species of them, nearly all of which have a different system of marching; eight were new to science when I sent them to England. Some are found commonly in every part of the country, and one is peculiar to the open campos of Santarém; but, as nearly all the species are found together at Ega, where the forest swarmed with their armies, I have left an account of the habits of the whole genus for

this part of my narrative. The *Ecitons* resemble, in their habits, the Driver-ants of Tropical Africa; but they have no close relationship with them in structure, and indeed belong to quite another sub-group of the ant tribe...

The peculiar feature in the habits of the *Eciton* genus is their hunting for prey in regular bodies, or armies. It is this which chiefly distinguishes them from the genus of common red stinging-ants (*Myrmica*), several species of which inhabit England, whose habit is to search for food in the usual irregular manner. All the *Ecitons* hunt in large organised bodies; but almost every species has its own special manner of hunting.

Eciton rapax. — One of the foragers, *Eciton rapax*, the giant of its genus, whose worker-majors are half-an-inch in length, hunts in single file through the forest. There is no division into classes amongst its workers, although the difference in size is very great, some being scarcely one-half the length of others. The head and jaws, however, are always of the same shape, and a gradation in size is presented from the largest to the smallest, so that all are able to take part in the common labours, of the colony. The chief employment of the species seems to be plundering the nests of a large and defenceless ant of another genus (*Formica*), whose mangled bodies I have often seen in their possession, as they were marching away. The armies of *Eciton rapax* are never very numerous.

Eciton legionis. — Another species, *E. legionis*, agrees with *E. rapax* in having workers not rigidly divisible into two classes; but it is much smaller in size, not differing greatly, in this respect, from our common English red ant (*Myrmica rubra*), which it also resembles in colour. The *Eciton legionis* lives in open places, and was seen only on the sandy campos of Santarém. The movements of its hosts were, therefore, much more easy to observe than those of all other kinds, which inhabit solely the densest thickets; its sting and bite, also, were less formidable than those of other species. The armies of *E. legionis* consist of many thousands of individuals, and move in rather broad columns. They are just as quick to break line, on being disturbed, and attack hurriedly and furiously any intruding object as the other *Ecitons*. The species is not a common one, and I seldom had good opportunities of watching its habits. The first time I saw an army, was one

evening near sunset. The column consisted of two trains of ants, moving in opposite directions; one train empty-handed, the other laden with the mangled remains of insects, chiefly larvae and pupae of other ants. I had no difficulty in tracing the line to the spot from which they were conveying their booty: this was a low thicket; the *Ecitons* were moving rapidly about a heap of dead leaves; but as the short tropical twilight was deepening rapidly, and I had no wish to be benighted on the lonely campos, I deferred further examination until the next day.

On the following morning, no trace of ants could be found near the place where I had seen them the preceding day, nor were there signs of insects of any description in the thicket; but at the distance of eighty or one hundred yards, I came upon the same army, engaged, evidently, on a razzia of a similar kind to that of the previous evening; but requiring other resources of their instinct, owing to the nature of the ground. They were eagerly occupied, on the face of an inclined bank of light earth, in excavating mines, whence, from a depth of eight or ten inches, they were extracting the bodies of a bulky species of ant, of the genus *Formica*. It was curious to see them crowding round the orifices of the mines, some assisting their comrades to lift out the bodies of the *Formicae*, and others tearing them in pieces, on account of their weight being too great for a single *Eciton*; a number of carriers seizing each a fragment, and carrying it off down the slope. On digging into the earth with a small trowel near the entrances of the mines, I found the nests of the *Formicae*, with grubs and cocoons, which the *Ecitons* were thus invading, at a depth of about eight inches from the surface. The eager freebooters rushed in as fast as I excavated, and seized the ants in my fingers as I picked them out, so that I had some difficulty in rescuing a few entire for specimens. In digging the numerous mines to get at their prey, the little *Ecitons* seemed to be divided into parties, one set excavating, and another set carrying away the grains of earth. When the shafts became rather deep, the mining parties had to climb up the sides each time they wished to cast out a pellet of earth; but their work was lightened for them by comrades, who stationed themselves at the mouth of the shaft, and relieved them of their burthens, carrying

the particles, with an appearance of foresight which quite staggered me, a sufficient distance from the edge of the hole to prevent them from rolling in again. All the work seemed thus to be performed by intelligent cooperation amongst the host of eager little creatures; but still there was not a rigid division of labour, for some of them, whose proceedings I watched, acted at one time as carriers of pellets, and at another as miners, and all shortly afterwards assumed the office of conveyors of the spoil.

In about two hours, all the nests of *Formicae* were rifled, though not completely, of their contents, and I turned towards the army of *Ecitons*, which were carrying away the mutilated remains. For some distance there were many separate lines of them moving along the slope of the bank; but a short distance off, these all converged, and then formed one close and broad column, which continued for some sixty or seventy yards, and terminated at one of those large termitariums already described in a former chapter as being constructed of a material as hard as stone. The broad and compact column of ants moved up the steep sides of the hillock in a continued stream; many, which had hitherto trotted along empty-handed, now turned to assist their comrades with their heavy loads, and the whole descended into a spacious gallery or mine, opening on the top of the termitarium. I did not try to reach the nest, which I supposed to lie at the bottom of the broad mine, and therefore in the middle of the base of the stony hillock.

Eciton drepanophora. — The commonest species of foraging ants are the *Eciton hamata* and *E. drepanophora*, two kinds which resemble each other so closely that it requires attentive examination to distinguish them; yet their armies never intermingle, although moving in the same woods and often crossing each other's tracks. The two classes of workers look, at first sight, quite distinct, on account of the wonderful amount of difference between the largest individuals of the one, and the smallest of the other. There are dwarfs not more than one-fifth of an inch in length, with small heads and jaws, and giants half an inch in length with monstrously enlarged head and jaws, all belonging to the same family. There is not, however, a distinct separation of classes, individuals existing which connect together the two extremes. These *Ecitons* are seen in the pathways of the

forest at all places on the banks of the Amazons, travelling in dense columns of countless thousands. One or other of them is sure to be met with in a woodland ramble, and it is to them probably, that the stories we read in books on South America apply, of ants clearing houses of vermin, although I heard of no instance of their entering houses, their ravages being confined to the thickest parts of the forest.

When the pedestrian falls in with a train of these ants, the first signal given him is a twittering and restless movement of small flocks of plain-coloured birds (ant-thrushes) in the jungle. If this be disregarded until he advances a few steps further, he is sure to fall into trouble, and find himself suddenly attacked by numbers of the ferocious little creatures. They swarm up his legs with incredible rapidity, each one driving its pincer-like jaws into his skin, and with the purchase thus obtained, doubling in its tail, and stinging with all its might. There is no course left but to run for it; if he is accompanied by natives they will be sure to give the alarm, crying "Tauoca!" and scampering at full speed to the other end of the column of ants. The tenacious insects who have secured themselves to his legs then have to be plucked off one by one, a task which is generally not accomplished without pulling them in twain, and leaving heads and jaws sticking in the wounds. The errand of the vast ant-armies is plunder, as in the case of *Eciton legionis*; but from their moving always amongst dense thickets, their proceedings are not so easy to observe as in that species. Wherever they move, the whole animal world is set in commotion, and every creature tries to get out of their way. But it is especially the various tribes of wingless insects that have cause for fear, such as heavy-bodied spiders, ants of other species, maggots, caterpillars, larvae of cockroaches and so forth, all of which live under fallen leaves, or in decaying wood. The *Ecitons* do not mount very high on trees, and therefore the nestlings of birds are not much incommoded by them. The mode of operation of these armies, which I ascertained only after long-continued observation, is as follows. The main column, from four to six deep, moves forward in a given direction, clearing the ground of all animal matter dead or alive, and throwing off here and there, a thinner column to forage for a short time on the flanks of the main

army, and re-enter it again after their task is accomplished. If some very rich place be encountered anywhere near the line of march, for example, a mass of rotten wood abounding in insect larvae, a delay takes place, and a very strong force of ants is concentrated upon it. The excited creatures search every cranny and tear in pieces all the large grubs they drag to light. It is curious to see them attack wasps' nests, which are sometimes built on low shrubs. They gnaw away the papery covering to get at the larvae, pupae, and newly-hatched wasps, and cut everything to tatters, regardless of the infuriated owners which are flying about them. In bearing off their spoil in fragments, the pieces are apportioned to the earners with some degree of regard to fairness of load: the dwarfs taking the smallest pieces, and the strongest fellows with small heads the heaviest portions. Sometimes two ants join together in carrying one piece, but the worker-majors with their unwieldy and distorted jaws, are incapacitated from taking any part in the labour. The armies never march far on a beaten path, but seem to prefer the entangled thickets where it is seldom possible to follow them. I have traced an army sometimes for half a mile or more, but was never able to find one that had finished its day's course and returned to its hive. Indeed, I never met with a hive; whenever the *Ecitons* were seen, they were always on the march.

The life of these *Ecitons* is not all work, for I frequently saw them very leisurely employed in a way that looked like recreation. When this happened, the place was always a sunny nook in the forest. The main column of the army and the branch columns, at these times, were in their ordinary relative positions; but, instead of pressing forward eagerly, and plundering right and left, they seemed to have been all smitten with a sudden fit of laziness. Some were walking slowly about, others were brushing their antennae with their fore-feet; but the drollest sight was their cleaning one another. Here and there an ant was seen stretching forth first one leg and then another, to be brushed or washed by one or more of its comrades, who performed the task by passing the limb between the jaws and the tongue, finishing by giving the antennae a friendly wipe. It was a curious spectacle, and one well calculated to increase one's amazement at the similarity between the

instinctive actions of ants and the acts of rational beings, a similarity which must have been brought about by two different processes of development of the primary qualities of mind. The actions of these ants looked like simple indulgence in idle amusement. Have these little creatures, then, an excess of energy beyond what is required for labours absolutely necessary to the welfare of their species, and do they thus expend it in mere sportiveness, like young lambs or kittens, or in idle whims like rational beings. It is probable that these hours of relaxation and cleaning may be indispensable to the effective performance of their harder labours, but whilst looking at them, the conclusion that the ants were engaged merely in play was irresistible...

The armies of all *Ecitons* are accompanied by small swarms of a kind of two-winged fly, the females of which have a very long ovipositor, and which belongs to the genus *Stylogaster* (family Conopsidae). These swarms hover with rapidly vibrating wings, at a height of a foot or less from the soil over which the *Ecitons* are moving, and occasionally one of the flies darts with great quickness towards the ground. I found they were not occupied in transfixing ants, although they have a long needle-shaped proboscis, which suggests that conclusion, but most probably in depositing their eggs in the soft bodies of insects, which the ants were driving away from their hiding-places. These eggs would hatch after the ants had placed their booty in their hive as food for their young. If this supposition be correct, the *Stylogaster* would offer a case of parasitism of quite a novel kind. Flies of the genus *Tachinus* exhibit a similar instinct, for they lie in wait near the entrances to bees' nests, and slip their eggs into the food which the deluded bees are in the act of conveying for their young.

Return to Pará

I arrived at Pará on the 17th of March, after an absence in the interior of seven years and a half. My old friends, English, American, and Brazilian, scarcely knew me again, but all gave me a very warm welcome, especially Mr. G. R. Brocklehurst (of the firm of R. Singlehurst and Co., the chief foreign merchants, who had been my correspondents), who received me into his house, and treated me with the utmost kindness. I was rather surprised at the

warm appreciation shown by many of the principal people of my labours; but, in fact, the interior of the country is still the "sertão" (wilderness), — a terra incognita to most residents of the seaport, — and a man who had spent seven and a half years in exploring it solely with scientific aims was somewhat of a curiosity. I found Pará greatly changed and improved. It was no longer the weedy, ruinous, village-looking place that it appeared when I first knew it in 1848. The population had been increased (to 20,000) by an influx of Portuguese, Madeiran, and German immigrants, and for many years past the provincial government had spent their considerable surplus revenue in beautifying the city. The streets, formerly unpaved or strewn with loose stones and sand, were now laid with concrete in a most complete manner; all the projecting masonry of the irregularly built houses had been cleared away, and the buildings made more uniform. Most of the dilapidated houses were replaced by handsome new edifices, having long and elegant balconies fronting the first floors, at an elevation of several feet above the roadway. The large, swampy squares had been drained, weeded, and planted with rows of almond and casuarina trees, so that they were now a great ornament to the city, instead of an eyesore as they formerly were. My old favourite road, the Monguba avenue, had been renovated and joined to many other magnificent rides lined with trees, which in a very few years had grown to a height sufficient to afford agreeable shade; one of these, the Estrada de São Jose, had been planted with coconut palms. Sixty public vehicles, light cabriolets (some built in Pará), now plied in the streets, increasing much the animation of the beautified squares, streets, and avenues.

I found also the habits of the people considerably changed. Many of the old religious holidays had declined in importance and given way to secular amusements; social parties, balls, music, billiards, and so forth. There was quite as much pleasure-seeking as formerly, but it was turned in a more rational direction, and the Paráenses seemed now to copy rather the customs of the northern nations of Europe, than those of the mother-country, Portugal. I was glad to see several new booksellers' shops, and also a fine edifice devoted to a reading-room supplied with periodicals, globes, and

maps, and a circulating library. There were now many printing-offices, and four daily newspapers. The health of the place had greatly improved since 1850, the year of the yellow fever, and Pará was now considered no longer dangerous to new comers.

Departure for England

June 2, 1859. — At length, on the second of June, I left Pará, probably forever; embarking in a North American trading-vessel, the *Frederick Demming*, for New York, the United States' route being the quickest as well as the pleasantest way of reaching England. My extensive private collections were divided into three portions and sent by three separate ships, to lessen the risk of loss of the whole. On the evening of the third of June, I took a last view of the glorious forest for which I had so much love, and to explore which I had devoted so many years. The saddest hours I ever recollect to have spent were those of the succeeding night when, the Mameluco pilot having left us free of the shoals and out of sight of land though within the mouth of the river at anchor waiting for the wind, I felt that the last link which connected me with the land of so many pleasing recollections was broken. The Paráenses, who are fully aware of the attractiveness of their country, have an alliterative proverb, "Quem vai para (o) Pará para," "He who goes to Pará stops there," and I had often thought I should myself have been added to the list of examples. The desire, however, of seeing again my parents and enjoying once more the rich pleasures of intellectual society, had succeeded in overcoming the attractions of a region which may be fittingly called a Naturalist's Paradise. During this last night on the Pará river, a crowd of unusual thoughts occupied my mind. Recollections of English climate, scenery, and modes of life came to me with a vividness I had never before experienced, during the eleven years of my absence. Pictures of startling clearness rose up of the gloomy winters, the long grey twilights, murky atmosphere, elongated shadows, chilly springs, and sloppy summers; of factory chimneys and crowds of grimy operatives, rung to work in early morning by factory bells; of union workhouses, confined rooms, artificial cares and slavish conventionalities. To live again amidst these dull scenes I

was quitting a country of perpetual summer, where my life had been spent like that of three-fourths of the people in gipsy fashion, on the endless streams or in the boundless forests. I was leaving the equator, where the well-balanced forces of Nature maintained a land-surface and climate that seemed to be typical of mundane order and beauty, to sail towards the North Pole, where lay my home under crepuscular skies somewhere about fifty-two degrees of latitude. It was natural to feel a little dismayed at the prospect of so great a change, but now, after three years of renewed experience of England, I find how incomparably superior is civilised life, where feelings, tastes, and intellect find abundant nourishment, to the spiritual sterility of half-savage existence, even if it were passed in the garden of Eden. What has struck me powerfully is the immeasurably greater diversity and interest of human character and social conditions in a single civilised nation, than in equatorial South America where three distinct races of man live together. The superiority of the bleak north to tropical regions however is only in their social aspect, for I hold to the opinion that although humanity can reach an advanced state of culture only by battling with the inclemencies of nature in high latitudes, it is under the equator alone that the perfect race of the future will attain to complete fruition of man's beautiful heritage, the earth.

The following day, having no wind, we drifted out of the mouth of the Pará with the current of fresh water that is poured from the mouth of the river, and in twenty-four hours advanced in this way seventy miles on our road. On the 6th of June, when in 7° 55' N. lat. and 52° 30' W. long., and therefore about 400 miles from the mouth of the main Amazons, we passed numerous patches of floating grass mingled with tree-trunks and withered foliage. Amongst these masses I espied many fruits of that peculiarly Amazonian tree the Ubussu palm; and this was the last I saw of the Great River.

Published in Great Britain by the Natural History Museum, Cromwell Road, London SW7 5BD

The introduction to this book draws upon the essay written by Maxwell Barclay for *Nature's Explorers*.

Reproduction by DL Imaging
Printed by Toppan Leefung Printing Ltd, China

Published in North America, South America, Central America, and the Caribbean by Smithsonian Books

This book may be purchased for educational, business, or sales promotional use. For information, please write: Special Markets Department, Smithsonian Books, P.O. Box 37012, MRC 513, Washington, DC 20013

ISBN 978-1-58834-687-2

Library of Congress Cataloging-in-Publication Data

Names: Bates, Henry Walter, 1825-1892, author, illustrator. | Bates, Henry Walter, 1825-1892. Naturalist on the river Amazons. | Bates, Henry Walter, 1825-1892. Insect fauna of the Amazon Valley.
Title: A naturalist in the Amazon : the journals & writings of Henry Walter Bates / Henry Walter Bates.
Description: London : The Natural History Museum, [2020] | Summary: "Facsimile reproductions of the drawings and notes of pioneering entomologist Henry Walter Bates documenting his 11-year-long travels in the Amazon in the mid-1850s"-- Provided by publisher.
Identifiers: LCCN 2019044276 | ISBN 9781588346872 (hardcover)
Subjects: LCSH: Natural history--Amazon River Valley. | Indians of South America--Amazon River Valley. | Insects--Amazon River Valley. | Amazon River Valley.
Classification: LCC QH112 .B368 2020 | DDC 508.81/1--dc23
LC record available at https://lccn.loc.gov/2019044276

Printed in China, not at government expense
24 23 22 21 20   1 2 3 4 5

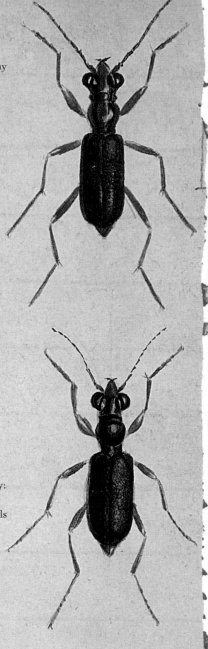